MatWerk

Edited by
Dr.-Ing. Frank O. R. Fischer (Deutsche Gesellschaft für Materialkunde e.V.)
Frankfurt am Main, Deutschland

Die inhaltliche Zielsetzung der Reihe ist es, das Fachgebiet „Materialwissenschaft und Werkstofftechnik" (kurz MatWerk) durch hervorragende Forschungsergebnisse bestmöglich abzubilden. Dabei versteht sich die Materialwissenschaft und Werkstofftechnik als Schlüsseldisziplin, die eine Vielzahl von Lösungen für gesellschaftlich relevante Herausforderungen bereitstellt, namentlich in den großen Zukunftsfeldern Energie, Klima- und Umweltschutz, Ressourcenschonung, Mobilität, Gesundheit, Sicherheit oder Kommunikation. Die aus der Materialwissenschaft gewonnenen Erkenntnisse ermöglichen die Herstellung technischer Werkstoffe mit neuen oder verbesserten Eigenschaften. Die Eigenschaften eines Bauteils sind von der Werkstoffauswahl, von der konstruktiven Gestaltung des Bauteils, dem Herstellungsprozess und den betrieblichen Beanspruchungen im Einsatz abhängig. Dies schließt den gesamten Lebenszyklus von Bauteilen bis zum Recycling oder zur stofflichen Weiterverwertung ein. Auch die Entwicklung völlig neuer Herstellungsverfahren zählt dazu. Ohne diese stetigen Forschungsergebnisse wäre ein kontinuierlicher Fortschritt zum Beispiel im Maschinenbau, im Automobilbau, in der Luftfahrtindustrie, in der chemischen Industrie, in Medizintechnik, in der Energietechnik, im Umweltschutz usw. nicht denkbar. Daher werden in der Reihe nur ausgewählte Dissertationen, Habilitationen und Sammelbände veröffentlicht. Ein Beirat aus namhaften Wissenschaftlern und Praktikern steht für die geprüfte Qualität der Ergebnisse. Die Reihe steht sowohl Nachwuchswissenschaftlern als auch etablierten Ingenieurwissenschaftlern offen.

It is the substantive aim of this academic series to optimally illustrate the scientific fields "material sciences and engineering" (MatWerk for short) by presenting outstanding research results. Material sciences and engineering consider themselves as key disciplines that provide a wide range of solutions for the challenges currently posed for society, particularly in such cutting-edge fields as energy, climate and environmental protection, sustainable use of resources, mobility, health, safety, or communication. The findings gained from material sciences enable the production of technical materials with new or enhanced properties. The properties of a structural component depend on the selected technical material, the constructive design of the component, the production process, and the operational load during use. This comprises the complete life cycle of structural components up to their recycling or re-use of the materials. It also includes the development of completely new production methods. It will only be possible to ensure a continuous progress, for example in engineering, automotive industry, aviation industry, chemical industry, medical engineering, energy technology, environment protection etc., by constantly gaining such research results. Therefore, only selected dissertations, habilitations, and collected works are published in this series. An advisory board consisting of renowned scientists and practitioners stands for the certified quality of the results. The series is open to early-stage researchers as well as to established engineering scientists.

Herausgeber/Editor:
Dr.-Ing. Frank O. R. Fischer (Deutsche Gesellschaft für Materialkunde e.V.)
Frankfurt am Main, Deutschland

Marcus Lau

Laser Fragmentation and Melting of Particles

With a Preface by Prof. habil. Dr.-Ing. Stephan Barcikowski

 Springer

Marcus Lau
Essen, Germany

Inaugural Dissertation, University of Duisburg-Essen, 2015

MatWerk
ISBN 978-3-658-14170-7   ISBN 978-3-658-14171-4 (eBook)
DOI 10.1007/978-3-658-14171-4

Library of Congress Control Number: 2016939382

Springer
© Springer Fachmedien Wiesbaden 2016

Printed on acid-free paper

This Springer imprint is published by Springer Nature
The registered company is Springer Fachmedien Wiesbaden GmbH

On the shoulders of giants

„Bernhard von Chartres sagte, wir seien gleichsam Zwerge, die auf den Schultern von Riesen sitzen, um mehr und Entfernteres als diese sehen zu können – freilich nicht dank eigener scharfer Sehkraft oder Körpergröße, sondern weil die Größe der Riesen uns emporhebt." – Johannes von Salisbury: Metalogicon 3,4,46-50 um 1159

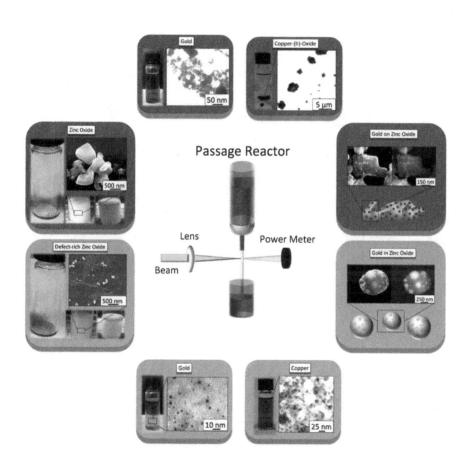

Preface

Particulate solids are attractive educts for the synthesis of colloids due to the availability of many materials as powders that can be handled occupationally safe as suspensions. Mechanical wet-grinding benefits from these advantages but bears drawbacks due to product contamination by grinding body abrasion. Further, especially hard or ductile educts are rarely accessible as nanoparticles by wet-grinding methods. Also the chemical modification is limited and can hardly be controlled by milling. The abrasion-free physico-chemical particle suspension processing by laser irradiation address these issues.

Prior to the work of Marcus Lau, presented here, several international studies on laser fragmentation and laser melting showed the general potential of the method. But all these studies could not extract data about the specific energy input or derive scaling parameters. These earlier studies focus on the final particle size and their composition, but are not investigating in detail the development of particle properties over time and the influence of laser and fluid parameters on the fabricated inorganic nanoparticle colloids. But exactly that is required for an additional understanding in this relatively new synthesis approach.

For this purpose Dr. Marcus Lau developed independently a comparable simple but useful apparatus and studied the influence of fluid, laser, and material parameters on the yield and properties of the particles irradiated with high power lasers. He exploits this new set-up for detailed studies of the process with selected model substances. This advanced technique enables a highly demanded quantification of the specific energy input, the passage quenching, and the identification of intermediates during the process, what gives insight into particle property development.

This book "Laser fragmentation and melting of particles" gives a general understandable introduction into the field of laser fragmentation and melting of particles. The presented state of the art summarizes the comparably young method of laser fragmentation of particles and the even younger method of laser melting of particles. Within the experimental studies he uses different educts as model substances such as metals (Au, Al), a semiconductor (ZnO), metal precursors (Cu_3N, CuO, Cu_2N, CuI) and a material combination (Au/ZnO) to show that mechanistic insights and technical relevant applications are possible. The investigations demonstrate that particle properties and the particle chemistry can be modified and yield-optimized by irradiation with high power, short and ultrashort pulse lasers.

The experimental work demonstrates that the developed passage reactor is suitable to control the specific energy input and energy density to manipulate the particle properties of different materials by laser irradiation. The different material characterization methods reveal that after each passage (irradiation cycle) particle properties can be adjusted with impressive accuracy. Furthermore, the experimental results and the modelling of the process demonstrate that this colloid synthesis method is close to overcome the lab scale. Additionally the results give insight into the possibilities to fabricate new materials. This is shown exemplarily for gold nanoparticles supported on zinc oxide particles that are successfully inverted resulting in inclusions of nanoparticulate gold in a zinc oxide matrix.

The dissertation of Marcus Lau is a pioneering work especially for the mechanistic understanding, to identify intermediates, and for upscaling for the laser irradiation process of particles in liquids. Enjoy reading, like I did.

Prof. habil. Dr.-Ing. Stephan Barcikowski

Content

List of Figures

List of Tables

Abbreviations

ADC	**A**nalytical **D**isc **C**entrifuge
AM	**A**dditive **M**anufacturing
BSE	**B**ack **S**cattered **E**lectron
Cw	**C**ontinuous **W**ave
DLS	**D**ynamic **L**ight **S**cattering
EELS	**E**lectron **E**nergy **L**oss **S**pectroscopy
FIB	**F**ocused **I**on **B**eam
GNS	**G**raphene **N**ano **S**heets
HR-TEM	**H**igh **R**esolution **T**ransmission **E**lectron **M**icroscope
IR	**I**nfra-**R**ed
ITO	**I**ndium **T**in **O**xide
Laser	**L**ight **A**mplification by **S**timulated **E**mission of **R**adiation
LMD	**L**aser **M**etal **D**eposition
MP	**M**icro**p**article
NP	**N**ano**p**article
Ns	**N**ano**s**econd
OB	**O**ptical **B**reakdown
PLAL	**P**ulsed **L**aser **A**blation in **L**iquids
PLFL	**P**ulsed **L**aser **F**ragmentation in **L**iquids
PLML	**P**ulsed **L**aser **M**elting in **L**iquids
PLPPL	**P**ulsed **L**aser **P**rocessing of **P**articles in **L**iquids
PI	**P**ost **I**rradiation
PPI	**P**rimary **P**article **I**ndex
Ps	**P**ico**s**econd
PSD	**P**article **S**ize **D**istribution
SDS	**S**odium **D**odecyl **S**ulphate
SE	**S**econdary **E**lectron
SEM	**S**canning **E**lectron **M**icroscopy

SLM	**S**elective **L**aser **M**elting
SLS	**S**elective **L**aser **S**intering
SMS	**S**ub-**m**icrometer **S**phere
TCO	**T**ransparent **C**onductive **O**xide
TEM	**T**ransmission **E**lectron **M**icroscope
UT	**U**n**t**reated
UV-vis	**U**ltra **V**iolet – **Vis**ible
XPS	**X**-Ray **P**hotoelectron **S**pectroscopy
XRD	**X**-**R**ay **D**iffraction

Physical parameters

Symbol	Notation	Unit
A_{com}	liquid jets cross section	[m²]
A_f	spot area at the front of liquids jet	[m²]
A_{ill}	illuminated area of liquid jets' cross section	[m²]
A_{un}	unilluminated area of liquid jets' cross section	[m²]
D	diameter of laser beam before focusing	[m]
d_b	focal spot diameter at liquid jets' back	[m]
d_f	focal spot diameter at liquid jets' front	[m]
d_p	Particle diameter	[m]
E_P	laser pulse energy	[J]
F	fluence	[J/cm²]
f	focal length of the lens	[m]
h	height of laser spot on liquid jets surface	[m]
I	intensity	[W/cm²]
K	beam quality number	[1]
M²	beam parameter product	[1]
$m_{P,ges}$	mass of particles in liquid	[g]
N_P	number of pulses per volume	[1]
n_P	number of particles	[1]
P_1	laser power before liquid jet	[W]
P_2	laser power behind liquid jet	[W]
$P_{j,b}$	position of liquid jets' back as distance to the focal plane	[m]

Symbol	Notation	Unit
$P_{j,f}$	position of liquid jets' front as distance to the focal plane	[m]
R_r	Repetition rate	[s^{-1}]
r_{ill}	ratio of illumination	[m^2/m^2]
V_c	volume of cone segment irradiated by laser beam	[m^3]
V_{sf}	volume of sphere segment at the front of the liquid jet	[m^3]
V_{sb}	volume of sphere segment at the back of the liquid jet	[m^3]
V_P	volume of a single particle	[m^3]
V_{com}	volume of irradiated area in the liquid jet by laser beam	[m^3]
v_l	Velocity of liquid	[m/s]
$\dot{V}$	volume flow rate	[mL/s]
$\dot{m}$	mass flow rate	[kg/s]
λ	Wavelength	[nm]
ρ_P	density of particles (bulk value)	[g/cm^3]
ρ_l	density of liquid	[g/cm^3]
$\tau_{e\text{-}p}$	electron-phonon coupling time	[s]
τ_P	laser pulse duration	[s]

1 Introduction

Even though nanoparticles were proposed to have manifold possibilities in application due to their unique properties, some materials and techniques remain inaccessible due to a missing link between synthesis and application or up-scaling. Thus, one of the challenging tasks in nanoscience, besides synthesis strategies remains bringing the nanostructures and particles to application. Top-down and bottom-up strategies can be differentiated, not only in the fabrication of nanomaterials, but also be considered for development of new nanomaterials including the required integration of nanomaterials along the processing chain. If a particular application is desired and properties of nanoparticles could perform this demand, the challenge is to bring the nanoparticles in a sufficient way onto or into the matrix of the nanofunctionalized material delivering function to the macroscopic product. Such a nanofunctionalization unfortunately is often the last step in a research and development chain where the properties of nanoparticles are accessed. Before developing such nanofunctionalized materials, properties of nanoparticles and the potential of utilizing them have to be explored and potential applications have to be encountered. Therefore, new nano-embedding synthesis strategies and precise process control is required to achieve reproducible product quality during the nanointegration chain. Furthermore, scalable techniques with defined process parameters correlating nanomaterial parameters will ease the way to application of nanomaterials and their continuous fabrication.

Here laser-generated nanoparticles are promising in a variety of fields such as biomedicine, optics or catalysis. Due to the possibility of receiving the nanoparticles without any ligands in a variety of liquids, even monomers or liquid polymers, from numerous materials there is potential for several applications in the stated fields.

Laser irradiation of educt particles in liquid environment enables chemical conversion and modification of particle sizes. This approach is straight forward if the solid educt material is in a particulate state anyway. Further, pulsed laser processing of particles in liquids (PLPPL) is a method that matches requirements for up-scaling because of its continuous process nature, but lacks of mechanistic understanding. PLPPL can be subdivided into two different processes. One is the pulsed laser fragmentation in liquids (PLFL) whereby particles are fragmentized resulting in a reduced particle size. The other processing technique of PLPPL is the pulsed laser melting of particles in liquid (PLML) where particles are transferred into a molten state and resolidify as spheres.

For PLPPL, a laser beam irradiates the particles that are suspended in a liquid that surrounds them. In the case of PLFL, these educt particles release the smaller product particles that are captured in the liquid. If the educt particles are exposed to laser parameters causing PLML, they melt, fuse and resolidify as spheres while the particle size either increases or remains constant. Whether PLFL or PLML occurs during PLPPL is depending on the laser parameters and the material response to incident light. PLFL can be classified as a wet comminution process. However, PLML is unique due to the fact that by irradiation with an electromagnetic wave the energy is directly transferred to the particle. Thus the particle can melt while it is in a liquid environment, or is fragmentized without any surfactants or impurities.

Within this work an experimental design that allows studying the influence of laser parameters is developed and demonstrated on obtained particle properties. PLPPL in a continuous liquid flow is studied whereby for PLFL as well as for PLML fluence regimes are shown. Precise control of laser fluence in the liquid jet enabled to study the effect of energy density on particle response causing PLFL or PLML. Thus changes of particle properties are correlated with defined laser parameters. The experimental design initially allows the study of the effect of laser fluence precisely enough to distinguish between different effects occurring during PLPPL. In order to reach sufficient energy density for both PLFL and PLML, especially in the case of metal oxide materials, the laser beam has to be focused. But this results in different energy densities during beam propagation behind the optical component focusing the laser beam. To this point, the laser beam has been always focused into a particle suspension vessel for PLPPL resulting in a strong deviation of laser fluence, depending on the plane in the liquid where the particle is irradiated. The unique experimental design used here addresses exactly this drawback and reduces the fluence deviation during laser irradiation of the particles moving in the liquid environment to a minimum. This firstly allows to study the impact of laser fluence for PLFL and PLML precisely enough to distinguish between different mechanisms occurring during PLPPL. Moreover the liquid jet method is prone to determine the transmitted laser energy behind the laser-excited educt particle suspension. Hence, a correlation of specific laser energy input with particle properties is possible. Since particle number and mass concentration is known to have an influence on PLFL the mass specific energy input and energy-per-particle values are accessible for a contribution to mechanistic understanding of PLPPL.

The developed passage reactor design enables quantitative particle processing and energy-specific efficiency modelling for experimental and scaling studies.

2 State of the art

Materials processing with lasers is an emerging field, within which a variety of laser processing techniques such as cutting, drilling, welding, piercing, patterning, rapid prototyping, macro- and micromachining, forming, cleaning and using lasers in biomedical processes have been established [Steen2010]. Laser processing of materials usually takes place in ambient air, thus using laser intensities that remove material from a workpiece's surface and the particles are released into the environment. These particles generated by laser irradiation are in the size regime of nanometers [Ullmann2002]. If these particles are released into ambient air, they bare the risk of adverse health effects when inhaled [Barcikowski2009a]. An alternative approach whereby particles are released yet captured is through laser materials processing in liquids [Fojtik1993]. After removal from the material's surface, the particles are confined in the liquid, thus avoiding their release into the workplace. This technique has been established in recent decades for nanoparticle generation, where it is not the processed workpiece that holds interest but rather the produced particles. A modification of this pulsed laser ablation in liquids (PLAL) technique is the laser irradiation of particles in the micro, sub-micro or nano range [Fojtik1993]. In contrast to PLAL, pulsed laser fragmentation in liquids (PLFL) is less intensely studied, resulting in the lack of mechanistic understanding and an awareness of the possibilities that this processing technique offers, which motivates this dissertation.

Chapters 2.1 and 2.2 will introduce to the materials selected as model materials for the practical part of this work, representing a semiconductor and metal material, respectively. Furthermore, an introduction is provided into the current state of the art to the interaction of particles with intense laser light in chapter 2.3, before the fundamental aspects of PLFL and PLML are explained (chapter 2.5 and 2.6). This will lead to the novel experimental set-up, which is developed and exploited, whereby an insight is provided into the possibilities that arise from the developed set-up. With this design, particle properties and laser parameters can be correlated and basic mechanistic conclusions of PLFL can be drawn. It will also be shown that for controlling PLFL, it is necessary to control the adapted laser fluence as precisely as possible. Furthermore, the developed set-up enables determining the specific energy input converted into particle properties.

2.1 Zinc oxide

Zinc oxide is commonly used as a color pigment, e.g. in paints [Wöll2007]. Moreover, it is also a semiconductor with a direct band gap of around 3.3 eV, a density of 5.61 g/cm³, a Mohs hardness of around 4.5 [Hernández Battez2008] and a wurtzite crystal lattice structure. It occurs rarely in nature and deposits can be found as zincite. After chemical synthesis, it forms a white powder. Özgür and Morkoç et al. reviewed the properties of zinc oxides and described pure zinc oxide, effects of doping and the variety of fabrication methods and application fields such as light-emitting devices, UV lasing, photodiodes, transparent conductive oxides (TCO), thin film transistors, piezoelectric devices, solar cells, gas sensors and bio sensors [Özgür2005], [Morkoç2008]. For several of these applications, engineering of the bandgap plays an important role in controlling and changing the electronic and photonic properties. Oba et al. showed that defects in the wurtzite ZnO lattice cause a change in the energy levels between the conductive and valence band [Oba2008]. Janotti et al. reviewed the possibilities of band gap engineering and effects of defects in zinc oxide, showing that n-type conductivity is promising for several applications [Janotti2008]. Dorranian et al. described photoluminescence of ZnO nanoparticles fabricated by PLAL from a Zn metal target in a water environment [Dorranian2012] attempting to correlate the photoluminescence properties to the applied wavelengths and fluences. Although they postulated an energy level scheme for the different possible transitions in the ZnO lattice, they could not clearly correlate generated particle properties with the laser parameters used for particle generation. Jadraque et al. showed the creation of oxygen vacancies on ZnO targets irradiated with different UV laser wavelengths under vacuum conditions [Jadraque2008]. For this, a 308 nm laser wavelength was more efficient to induce these defects compared to 266 nm. This shows that bandgap engineering and control of zinc oxide properties is a relevant research field due to the potential in application of different modifications from zinc oxide, such as TCO [Janotti2008]. Owing to the comparably low price and basic availability, it is a potential substitute for ITO [Ellmer2011]. Hiramatsu et al. showed that highly conductive TCO films can be generated by laser ablation of target materials in gas phase and a subsequent deposition onto a substrate [Hiramatsu1998]. They ablated target materials prepared from zinc oxide powders containing germanium oxide or aluminum oxide powders, known to be an effective dopant for this purpose [Luo2013]. For this, the powders were mixed and sintered prior to laser ablation, which resulted in a highly conductive TCO onto the substrate, comprising the comparably cheap educts such as ZnO and Al_2O_3 powders. Usui et al. reported about the laser ablation of a zinc plate and obtained zinc oxide nanoparticles [Usui2005], thus showing the (partial) oxidation of the material. Nonetheless, a comprehensive understanding is lacking [Avadhut2012], [Ciupina2004], [Yang2012], [Rajeswari2011], [Kelchtermans2013].

Further engineering of zinc oxides bandgap is a promising method to modify the material. This holds particular interest as zinc oxide and defect-rich zinc oxide particles are promising, e.g. for catalytic applications [Wöll2007]. Lin et al. determined green emission from oxide antisite defect rather than a crystallographic vacancy or interstitial [Lin2001]. They proposed the scheme in Figure 1 for energy levels [Lin2001], whereby this band gap (E_g) results in a local peak around 370 nm for ZnO nanoparticle dispersions in the UV-vis extinction spectrum [Srikant1998], [Zak2011].

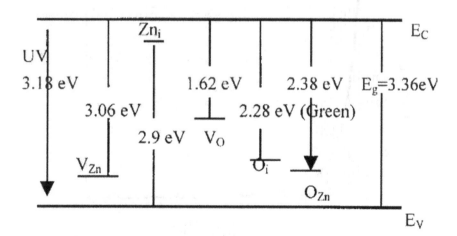

Figure 1: Scheme of energy levels with energy levels of conduction band (E_C), valence band (E_V), zinc vacancy (V_{Zn}), interstitial zinc (Zn_i), oxygen vacancy (V_O), interstitial oxygen (O_i) and oxygen antisite defect (O_{Zn}) (taken with permission from Lin et al. [Lin2001])

For processing of zinc oxide particles in liquid environment, its solubility also needs to be considered. Schindler et al. reported about the free enthalpy of formation [Schindler1964] and solubility in water in dependence of zinc oxide's particle size [Schindler1965]. Besides, ZnO, $Zn(OH)_2$, α-$Zn(OH)_2$, β_1-$Zn(OH)_2$, β_1-$Zn(OH)_2$, γ-$Zn(OH)_2$, δ-$Zn(OH)_2$, and ϵ-$Zn(OH)_2$ can occur as solid states in a liquid environment [Schindler1964].

Schindler et al. also proposed the following equation for zinc oxides solubility in dependence of the particle diameter [Schindler1965]:

$$\log Ks_0 = -16(\pm 0.05) + \frac{50(\pm 26)}{d_p \cdot 10}$$

With Ks_0: solubility product, d_p: particle diameter in nm

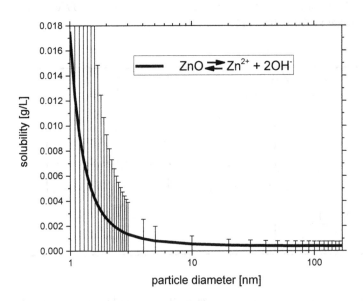

Figure 2: Calculated solubility of ZnO in water at a pH of 6.7 in dependence of the particle diameter [Schindler1965]

Figure 2 shows a diagram for the solubility plotted versus the particle diameter using the equation of Schindler et al. with the corresponding equation for dissolution of ZnO in water environment. For particle diameters below 5 nm, this solubility significantly increases but does not exceed 18 mg/L for 1 nm particles. Note that the error bars are larger; thus, the solubility of such small particles might be higher.

The solubility of ZnO powder from the GESTRIS data base is given with 1.6 mg/L, meaning that zinc oxide is almost not soluble in water.

To summarize, it can be stated that zinc oxide is promising for several applications besides its use as a white pigment in simple paints, due to the manifold electronic structure [Lin2001]. Thus, it can be used, in catalysis [Wöll2007] or as varistors in electronic devices [Gupta1990], for instance. Several potential applications arise when it is in the nanoscale [Wang2004], such as a nanolasing device [Huang2001].

Accordingly a method to modify zinc oxides electronic structure and size by a precisely controlled process might be promising.

2.2 Gold nanoparticles

Gold nanoparticles (Au NP) hold particular interest in a variety of fields such as catalysis [Hartua1997], as well as biology and medicine [Giljohann2010], [Murphy2008]. The possible application fields and different synthesis strategies have been studied intensely over many decades [Daniel2004]. Due to their optical properties which differ from the bulk material and are size- and shape-dependent the interest in Au NP ranges from fundamental research to real-world applications. The discovery of deviation of optical properties for gold nanoparticles compared to the bulk material dates back to Faraday in 1857 [Faraday1857].

Many chemical bottom-up synthesis routes exist for the fabrication of gold nanoparticles, whereas only a few top-down methods provide access to Au NP. One of the most common synthesis routes for gold nanoparticles was reported by Turkevich in 1951 [Turkevich1951]. Another popular bottom-up synthesis is the Brust-Shiffrin method [Brust1994]. Turkevich's citrate-based method delivers particle sizes between 10-25 nm that show good colloidal stability in water. The advantage of Brust's method to obtain comparably small particle sizes and widths of particle size distributions is accompanied by the drawback that the thiols used for size quenching strongly bind to particles' surface [Häkkinen2012], [Xue2014], thus making a subsequent ligand exchange or removal difficult. Overall, a variety of synthesis routes and potential applications are discussed and shown, although for some applications (e.g. biological, catalytic) impurities like residual chemicals or abrasion from comminution processes are an immense problem [Goesmann2010], [Raab2011]; thus, subsequent cleaning steps are required.

A real-world application in which purity and size plays an important role is the use of gold nanoparticles in catalysis [Haruta1997]. Masatake Haruta first showed the activity of gold nanoparticles in catalysis and its size dependency on the activity [Haruta1987], [Haruta1997]. For catalytic applications, small particles are preferential. Activity of gold nanoparticles in catalysis is correlated to the free surface and significant increase in activity can be observed below 5 nm [Haruta1997]. As support particles for gold nanoparticles, titanium dioxide is used, e.g. for oxidation of CO (water-gas shift reactions) [Sakurai1997].

Here, the laser generation of gold nanoparticles comes into play, which can deliver pure and ligand-free gold nanoparticles harvested in a liquid environment. A synthesis method bypassing the drawback of a ligand-occupied surface is laser ablation in liquids. First reported by Fojtik and Henglein [Fojtik1993], this method has gained increasing interest as many efforts in up-scaling and process control have been established in the last decade. The PLAL technique fabricates ligand-free gold nanoparticles from a plasma plume, which is followed by cavitation bubbles directly from the bulk material [Ibrahimkutty2012]. Besides the expansion and collapse of one first major cavitation bubble which is followed by smaller ones that expand and collapse on the target's surface it could be shown that primary particles in the size range of around 7 nm can pass the phase boundary, whereas larger aggregates and agglomerates with 40-60 nm are kept in the bubble [Wagener2013]. The detailed small angle X-ray scattering experiments by Wagener et al. provided an insight into what happens inside the cavitation bubble and where and when particles are formed. Beside this, there are also investigations and evidence of metal atom clusters formed after PLAL [Giorgetti2014]; nonetheless, PLAL generally delivers a wide particle size distribution. Recently Rehbock et al. reviewed how monodisperse and ligand-free gold nanoparticles can be obtained by PLAL [Rehbock2014].

Due to the possibilities in manufacturing ligand-free particles by PLAL e.g. freedom of material and solvent [Baersch2009], embedding them into polymers for biological applications [Hahn2010], [Sowa-Soehle2013] or supporting the nanoparticles to microparticle supports [Wagener2012a] this technique has attracted increasing interest and use [Barcikowski2009], [Asahi2015].

Given that catalysis application requires free surfaces and small particle diameter, ultra-small ligand-free particles would be desirable. However, unfortunately to prevent these small gold nanoparticles from growing when they are synthesized e.g. by chemical reduction strong binding organic ligands are necessary [Schmid1981], [Schmid2008]. These ligands enable precise control of ultra-small gold particle sizes but simultaneously cover the surface.

The size limit for laser-generated and ligand-free gold nanoparticles is reported at around 4 nm in diameter [Amendola2007], [Rehbock2014], although smaller ligand-free nanoparticles would be desired due to higher specific surface area and probably high reactivity.

2.3 Different lasers for one application – particle processing

Since the first report of light amplification by stimulated emission of radiation (laser) in ruby by Maiman [Maiman1960a], [Maiman1960b], [Maiman1962] many efforts and developments have been engaged in manufacturing lasers with higher intensity. The following figure depicts the development of the available focused laser intensity versus time (left) and the peak power plotted versus the average power for different femtosecond systems (right) [Sibbett2012]. The diagram from Sibbett et al. also indicates the average and peak power demanded for different applications such as biomedicine, telecommunications and materials processing, highest demands in average power and pulse peak power are found for materials processing, including laser ablation.

At present, these high intensities are only achieved by pulsing of the laser. The duration of a single released pulse (pulse length) has a crucial impact on the response of the processed material regarding short (nanoseconds) and ultra-short (starting from femtoseconds to a few picoseconds) pulsed lasers, owing to the relaxation time of electrons releasing their energy to the lattice structure of atoms in solids in the range of picoseconds [Chichkov1996]. Due to this relaxation time, materials processing with picosecond lasers can be regarded as cold processes, whereas nanosecond materials processing will cause significant thermal effects to the material. In this work, all thresholds are described as fluence at fixed pulse duration defined as the energy of a single pulse divided by the area excited by focusing the laser beam. However, these thresholds only hold for the applied pulse length. Therefore, laser fragmentation with picosecond pulses will have lower fragmentation threshold fluences compared to nanosecond pulses, although the energy dose put into a particle is the same, owing to the aforementioned relaxation time of electrons.

It could be estimated that dividing the pulse energy by its duration (pulse length) will give an intensity threshold for material processing (depending on the optical material properties) that is independent on pulse length. However, due to the difference in mechanisms occurring for different pulse length for material processing, a uniform fragmentation (or ablation) threshold with the unit "intensity per pulse" which is independent of the pulse length will not exist. Nonetheless, as the mechanism occurring for material ablation or disruption can be correlated to the electron-phonon coupling time, there should be a possible differentiation in case of undergoing this electron-phonon coupling time and for sufficient time for electron relaxation (heating of the material). Lin and Zhigilei reported this time for Au in the range of 6 ps to 20 ps [Lin2003] and for ZnO this value is reported to be around 500 fs, thus half a picosecond [Zhukov2012]. This means that for ZnO including the case of the 10 ps used within the experiments there is sufficient time to cause thermal effects. However, for Au, the transition regime is addressed with 10 ps, which might make thermal effects e.g. melting or heating-evaporation more difficult compared to ZnO.

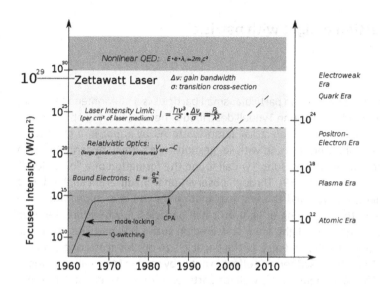

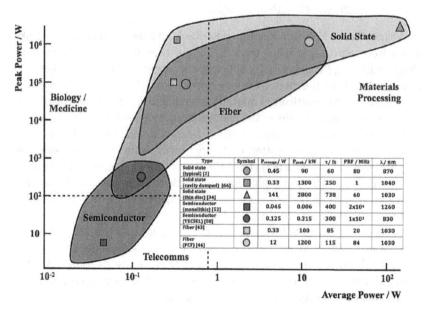

Figure 3: Development of available focused laser intensity over time [published work into the public domain at wikimedia commons, Wikipedia 03/21/2015] (top) and peak power versus the average power for different femtosecond systems taken with permission from Sibbett et al. [Sibbett2012] (bottom)

2.4 Interaction of light with particles

Light scattering

Light scattering of particles in particular small particles is a phenomenon that has long fascinated scientists. John Tyndall described general scattering observations of condensed light from an electrical lamp scattered by vapor in a glass tube. Inspired by Sir John Herschel and with suggestions of a refined experiment from Prof. George Gabriel Stokes, John Tyndall found the formation of a "blue cloud" in the glass tube [Tyndall1869]. Through observation from different angles during his experiment, he provided the first explanation for the blue appearance of a cloud-free sky, the scattering of light on small particles. Furthermore, he even stated the polarization of light when scattered on such small particles. Although he also mentioned a dependence of the particle size on this scattering effect, it was Gustav Mie who described this effect in detail by solving Maxwell's equations [Mie1908]. The Mie-Theory holds well for particles with diameters of 2-10 times λ, whereby λ is the wavelength of the light.

John William Strutt, 3rd Baron Rayleigh described the blue appearance of the sky and sea from particles smaller than 0.2 λ and thus this effect is known as Rayleigh-scattering [Rayleigh1910], although the meteorologist John Aitken had previously described the same phenomenon [Aitken1881]. John Aitken found that "extremely small particles of matter suspended in water" [Aitken1881] only scatter the short wavelengths of the light (although Rayleigh first described the size dependency mathematically). From his experiments, he proved that the hitherto-named, "selective scattering theory" holds rather than the "selective absorption theory" [Aitken1881]. The former stated that scattering of the blue region of light waves is responsible for the bluish sea color and the latter that selective absorption of the red color region is responsible for this appearance. This shows that even over one hundred years ago, scientists were familiar with the different possibilities of light and matter interaction and the dependence of the light wavelengths. Indeed, even nowadays, the optical appearance of the sky remains a relevant research topic [Gedzelman2005]. Rayleigh found that the intensity of scattering is proportional to the volume of a particle and proportional to λ^{-4} of the incident light wavelength [Rayleigh1899].

A drawback during the experimental observations during that time was that no monochromatic light source was available. Owing to the dependence of the incident angle of light and the wavelength of the light on the scattering effect on a particle with a defined size, a monochromatic light source with parallel beam with no or low divergence would be favorable. Thus, the discovery of light amplification by stimulated emission of radiation (LASER) was beneficial for further investigations [Maiman1960a], [Maiman1960b]. In particular, the availability of increasingly higher laser energies offers new possibilities in particle excitation.

Light absorption

The portion of light that is neither scattered nor transmitted through the particle is absorbed. Sufficient and fast excitation results in the formation of electron hole pairs within a few femtoseconds [Amendola2008]. These are transferred into hot electrons during a few hundreds of femtoseconds. Based upon the two temperature model, energy or heat is transferred to the atom lattice within a few picoseconds [Chichkov1996]. As a result, the excited electrons transfer energy in the form of heating of the atomic lattice occurs within a few picoseconds, e.g. for gold [Lin2003], [Werner2011a]. Note that if energy input is caused with ultra-short pulses fragmentation mechanisms will differ, whereby a so-called coulomb explosion is known to cause a cold ablation of material [Chichkov1996].

Subsequently, energy transfer from electrons to atoms causes heating of the particle. The resulting temperature can be determined from the energy portion transferred into the particle although the temperature dependency of the specific heat capacity should be considered [Furukawa1968]. If a solid particle in liquid environment heats up, four temperature regimes should be distinguished. If the time of particle heating is much shorter than the time required for sufficient heat transfer (from heated particle to the surrounding medium), it is possible that a liquid or vaporized particle is directly surrounded by the liquid medium for a short period of time. As a result, the molten or vaporized material can dissolve (e.g. as ions or atoms) directly in the liquid environment if soluble in the present concentration. For sufficient heat transfer, the system can undergo the following stages during particle heating and energy transfer to its liquid environment: solid particle surrounded by liquid, solid particle surrounded by gas, liquid particle surrounded by gas, vaporized particle surrounded by gas. Considering the possibility of a shock wave-induced particle fragmentation [Zhigilei1998], the possible stages shown in Figure 3 might occur after intense laser light absorption for a single particle.

This holds particular interest as this work addresses the intense irradiation with pico- and nanosecond laser sources. Figure 3 sketches the possible excitation and relaxation stages based upon literature and own findings. Vaporization [Schaumberg2014], bubble formation [Hashimoto2012], shock wave-induced fragmentation [Zhigilei1998] and particle melting [Link1999a] are known particle responses to intense photonic excitation.

After sufficient laser pulse absorption (see Figure 3 top) three different time regimes can be distinguished. For τ_P (duration of a single pulse) >> τ_{e-p} (material specific electron-phonon coupling time), electrons have sufficient time to transfer the energy received from laser irradiation (phonons) to the atomic lattice. Thus, they can transfer the energy resulting in heating, melting and vaporization of the particle for sufficient energy input. These stages potentially occurring for τ_P >> τ_{e-p} in chronological order (from top to bottom) are shown in the left column of Figure 3 [Link1999a], [Hashimoto2012], [Schaumberg2014].

For $\tau_P \approx \tau_{e-p}$, it might be estimated that both effects can occur, namely the mechanism of shock wave-driven particle fragmentation and heating-melting-evaporation. The electron-phonon coupling times for zinc oxide (around 0.5 ps) [Zhukov2012] and gold (around 6 to 20 ps) [Lin2003] fall in the range of 1-10 ps lasers. As τ_{e-p} time for Au is longer than τ_{e-p} time for ZnO, it might be estimated that with 10 ps laser pulses ZnO might be more effectively heat affected [Lau2014a]. This is schematically illustrated in the center column of Figure 3.

A distinctive shock wave-driven fragmentation mechanism might be assumed for τ_P << τ_{e-p} [Zhigiliei1998]. If the laser pulse duration is distinctively lower that the material dependent τ_{e-p}, electrons can be sufficiently excited to cause a coulomb explosion of the material without being heat affected [Chichkov1996]. This is illustrated in the column on the right of Figure 3.

Note that these are only suggested relaxation routes based upon the reported observations and theory from literature, although this is in agreement considering the electron-phonon coupling times for zinc oxide [Zhukov2012], and with own experimental findings as will be shown later.

photonic excitation

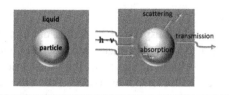

possible relaxation routes

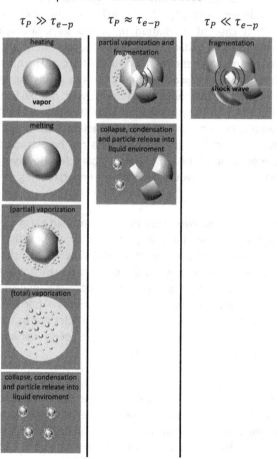

Figure 4: Proposed possible relaxation routes of a small particle after intense laser light absorption

2.5 Fundamentals of laser fragmentation

Intense laser light excitation of particles suspended in liquids enables their modification in structure and size. If the target particles are irradiated with a sufficient energy dose, their size can decrease by one of the pathways described in chapter 2.4, resulting in the formation of nanoparticles. This fabrication method of nanoparticles from suspended micro- or nanoparticles is mostly described as laser fragmentation. Besides the term "laser fragmentation", "laser ablation (of particles)" [Schaumberg2014], [Jeon2007], [Asahi2008] or "(post)- irradiation" ("post-" in case of irradiation of previously fabricated nanoparticles by PLAL) [Amendola2007], [Zhang2003] can be found in literature, although these terms usually do not match the mechanisms that occur. The term laser ablation (of particles) results from a problem that invariably arises for all particle size reduction techniques. For all comminution processes, the particle size distribution changes and this is usually associated with an educt-product particle mixing. Thus, investigation of the smallest fraction after laser fragmentation is most convenient, because the nanoparticles can be extracted from educt particles by sedimentation of the educt (micro-) particles. However, an analysis of the whole particle fractions would allow better understanding the PLFL process.

This work will show that PLFL can follow two major mechanisms, namely shock wave-induced particle disruption and nanoparticle generation by vaporization. The size difference between the educt MP and the NP fabricated by vaporization enables the separation of these two fractions. As a result, some researchers name this process laser ablation of MPs, neglecting the shock wave-induced particle fragmentation. Therefore, they do not consider all fragments of particles that are generated, such as shock wave-fragmentized microparticles not in the nano-sized regime. These fragments may represent the largest mass fraction, whereby the entire process thus far is only partially regarded.

To gain further insights into the mechanisms of laser fragmentation, experiments with controlled laser fluence are required. In this work, an experimental design will be demonstrated that allows studying mechanisms of laser fragmentation without the drawback of strong laser fluence deviation. This will be shown and discussed in the experimental section, where comparison to the state of the art is also explored. In general, interaction of laser light with metallic (plasmonic) and dielectric materials has to be differentiated: the former has been studied intensely using gold nanoparticles and several mechanisms depending on applied laser parameters have been described [Werner2011], [Werner2011a], [Katayama2014], [Hashimoto2012], [Link1999c], whereas for the latter only few theoretical and practical studies have proposed mechanisms for the laser-induced fragmentation [Zhigilei1998], [Cai1998]. Due to the difference in their electronic structure, metals and metal oxides are proposed to interact differently to laser irradiation. Therefore, the interaction of plasmonic and non-plasmonic materials will be presented in two different sections.

Precise studies determining the fluence thresholds for non-metallic particles do not exist. Heise et al. reported a strong dependence of the substrate for the ablation threshold of ZnO layers (around 1 µm) [Heise2011]. In case of a transparent glass substrate, they determined the ablation threshold to be around 4.5 (±0.5) J/cm² for 532 nm wavelength and 10 ps pulse length [Heise2011]. In case of an absorbing substrate, the threshold was found to be around 0.024 (±0.005) J/cm². Risch et al. reported the ablation threshold of 200 nm gallium doped ZnO layers on glass substrates. They found for 532 nm thresholds of around 1.1 J/cm² or 0.84 J/cm² for rear side laser irradiation [Risch2011]. The deviation to the findings of Heise et al. might arise from the doping of ZnO with gallium.

For plasmonic nanoparticles, Pyatenko et al. simulated the fluence thresholds for different laser wavelengths of nanosecond lasers (fundamental, 2nd harmonic and 3rd harmonic of Nd:YAG lasers) in dependence of the particle size [Pyatenko2013]. This valuable review covers silver, gold, copper, platinum and palladium nanoparticles and provides comprehensive insights. Together with the mechanistic findings from Hashimoto and co-workers [Hashimoto2012], a detailed understanding of the mechanisms and thresholds can be drawn. The mechanistic understanding of PLFL for non-metallic and metallic particles will be discussed in 2.5.1 and 2.5.2, respectively.

2.5.1 Interaction of laser light with non-metallic particles

At present, only little is known about the fragmentation mechanisms for ceramic or metal oxide particles interacting with intense laser light. Theoretical studies investigating the interaction of inorganic and organic particles were reported in 1997 and 1999, respectively [Cai1998], [Zhigilei1998], whereby their simulations revealed a particle disruption from the inside. Studies of Yeh et al. showed the partial reduction of CuO to elemental Cu nanoparticles in isopropanol [Yeh1998] [Yeh1999], thus changing the nanoparticle composition by reduction of copper-(II)-oxide microparticles partially to elemental copper nanoparticles. Interestingly, they obtained higher colloidal stability of their elemental copper nanoparticles than reported for chemically synthetized [Yeh1999]. In a subsequent study, Schaumberg et al. reported the formation of elemental copper nanoparticles from copper compounds such as CuO, Cu_3N and Cu_2C_2 in ethyl acetate [Schaumberg2014]. The results for copper nanoparticle formation from CuO of these practical studies indicate a material vaporization and recomposition under reductive conditions. Thus, the term ablation from particles is used by some researchers as the mechanism might be similar to PLAL. Considering the studies mentioned, it is obvious that there should be different mechanisms for particle size reduction, such as a disruption which might be driven by shock waves [Cai1998], [Wagener2012] and a material vaporization similar to PLAL. However, reports demonstrating that both of these mechanisms simultaneously occur, do not exist, to my best knowledge.

Beside these findings, fragmentation studies of organic materials showed that especially for ultra-short laser pulses (femtosecond) the particle decomposition is less significant than for nanosecond pulses [Sylvestre2011]. The variation of laser power for nanosecond and femtosecond PLFL revealed smaller particles with higher laser power as well as a higher degree of degradation of the pharmaceutical substrate. This might already be an indication of the two mechanisms simultaneously occurring for organic particles, namely disruption or fragmentation and vaporization. Extensive studies of fragmentation of organic microparticles are reported by the Asahi group [Tamaki2000], [Tamaki2002], [Sugiyama2006], [Jeon2007], [Asahi2008], [Sygiyama2011], who showed the (partial) preservation of organic particle composition after laser irradiation with nanosecond lasers using a variety of organic materials and solvents. In contrast to metallic and organic materials, mechanistic understanding of laser fragmentation for inorganic and non-metallic particles is poor despite these materials being intensively investigated by laser ablation, already focusing of breaking the limit of gram per hour yield [Sajti2010], [Intartaglia2014], [Wagener2010]. This limitation for PLAL results from shielding effects of the plasma-induced cavitation bubble hampering the subsequent laser pulse to reach the target. To avoid this and achieve a high productivity, high scanning velocities and repetition rates are necessary [Wagener2010]. Thus, PLFL of these materials will also hold interest because laser fragmentation might be more efficient than

PLAL. This assumption can be driven from a higher degree of freedom that suspended particles have compared to a bulky target material. Studies for PLAL of thin wires have shown an improved productivity [Messina2013]. For this purpose, one possible explanation is the improved specific surface area that can be reached by the focused laser beam. If the laser beam is focused at the tip of a thin metal wire, the nanoparticles can be released from the cavitation bubble, which can expand in almost all directions into the liquid. The only confinement is the longitudinal elongation of the wire. This elongation does not exist for particles completely covered by the laser beam. Note that there are no studies or evidence of cavitation bubbles reducing the process effectivity of PLFL. Moreover, studies reporting the nanoparticle yield per pulse from PLFL remain absent. A study by Jang et al. investigated the shock-wave formation at Cu microparticles in gas phase and determined velocities from 1000-4000 m/s and lower thresholds for ablation compared to a Cu bulk target [Jang2004]. This is another good indication for a possibly higher yield of PLFL compared to PLAL.

Considering the aforementioned investigations on laser fragmentation, it is obvious that the influence of pulse length [Sylvestre2011], laser wavelength, applied laser fluence [Sylvestre2011] and educt particle properties does not reveal a comprehensive laser fragmentation model for non-metallic particles at present. Furthermore, it is unclear when particle composition will be preserved for nanoparticles and when the composition changes. Preservation of material composition as well as decomposition or change in stoichiometry has been shown for nanosecond lasers, as discussed in the aforementioned studies by Yeh et al. [Yeh1998], [Yeh1999], Sylvestre et al. [Sylvestre2011] and Tamaki et al. [Tamaki2000], [Tamaki2002]. Own preliminary investigations on zinc oxide particles showed that nanoparticle formation efficiency could be increased by pre-treatment of the microparticles in a stirred media mill, resulting in the activation of the microparticles fabricating spherical nanoparticles (Fig. 5) [Wagener2012]. From these findings, a contribution of a shock wave and laser-induced plasma at particles' surface for PLFL contribution was assumed. Considering the state of the art, there is a demand to understand the contribution of laser fluence to PLFL for non-metallic particles, which will be addressed accordingly in chapters 4.1, 4.2, and 4.5 of the experimental sections.

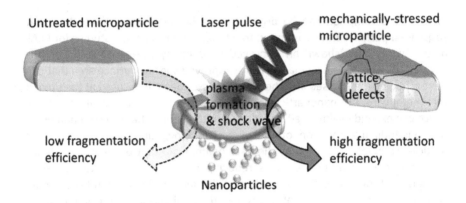

Figure 5: Schematic illustration of nanoparticle release from microparticles found for zinc oxide taken with permission from [Wagener2012]

2.5.2 Interaction of laser light with metallic particles

Some of the earliest studies on laser fragmentation of colloidal metallic nanoparticles date back to 1999, when Link and El Sayed reported that the second harmonic of Nd:YAG lasers fits well to surface plasmon resonance of gold nanoparticles [Link1999a], [Link1999b]. Their experiments also covered investigations of laser pulse length and fluence [Link1999c]. They addressed a wide regime of pulse length and laser fluence at the same time. Link et al. could show that laser treatment of nanorods results in sequential formation of spherical particles for different fluences and pulse lengths at 800 nm wavelength. This wavelength corresponds to the 2nd plamon peak of the nanorods used and thus was selective for rods' excitation. Due to a stepwise transformation from Au rods to spheres, these investigations might be considered as a first approach and strategy enabling energy dose balancing. Following these reports, a variety of plasmonic materials were investigated for their response to laser irradiation [Amendola2009], showing that laser parameters such as laser fluence can be correlated to obtained nanoparticle properties such as diameter. Mafuné and Kondow reported about small platinum nanoparticles stabilized by SDS generated with 355 nm laser irradiation [Mafuné2004] as well as gold nanoparticles stabilized with SDS for PLFL with 532 nm laser wavelength [Mafuné2001]. This shows that besides the laser wavelength and fluence, the surfactant concentration also has a crucial impact on received nanoparticle sizes. The crucial role of surface chemistry in particular oxidation for Au NP produced by PLAL in water was first described by Sylvestre et al. [Sylvestre2004], [Sylvestre2004a].

Alloy formation by laser irradiation of mixed Ag-Au particle microparticle suspensions was already reported in 2003 by Zhang et al. [Zhang2003], who adapted the idea of alloying nanoparticles by laser irradiation from Chen et al. [Chen2001]. These studies mentioned the advantages of using educt microparticles for alloy formation compared to alloy formation from colloidal mixtures of nanoparticles, although PLAL might be considered as the method of choice for alloy formation to avoid educt-product mixing. Regarding this issue, Jakobi et al. reported the preservation of stoichiometry after ablation [Jakobi2011] and Neumeister et al. characterized crystalline and homogenous ultra-structures of the produced alloy NPs [Neumeister2014]. Furthermore, Neumeister et al. cross-checked the alloy formation by PLAL with post-irradiation of colloidal mixtures from Au and Ag nanoparticles, revealing no alloy formation.

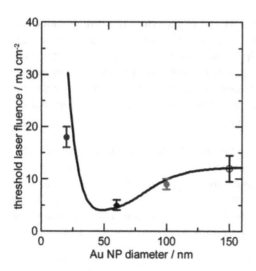

Figure 6: Bubble formation threshold for different gold nanoparticle diameter excited with 15 ps pulses and 355 nm wavelength [Katayama2014]; Reprinted (adapted) with permission from [Katayama2014]; Copyright (2014) American Chemical Society

The interaction of particles with the incident laser light strongly depends on particles extinction cross section and their absorption efficiency at the laser wavelength used [Link1999a]. Katayama et al. investigated the threshold for the formation of a bubble that is generated around plasmonic particles by laser irradiation in dependence of nanoparticle size. Figure 6 depicts the curve showing this fluence threshold for different particle diameter using an excitation pulse length of 15 ps [Katayama2014].

The findings for bubble formation threshold (Fig. 6) correlate with experimental findings for fragmentation thresholds from Cavicchi et al., who used 532 nm wavelength and 7 ns pulse length for the change of gold nanoparticle sizes for different laser fluences and educt particle diameter [Cavicchi2013]. Therefore, independent of pulse length and wavelength, particle sizes between 30 to 100 nm exhibit most efficient laser light absorption and lowest bubble formation thresholds [Katayama2014], [Pyatenko2013] [Gökce2015a]. In particular, studies by Cavicchi et al. revealed the lowest fragmentation threshold for 60 nm Au NP comparing 10nm, 20nm, 30 nm, 60 nm and 100 nm Au NP sizes [Cavicchi2013], indicating a correlation between bubble formation threshold and the fragmentation threshold.

Due to this reason, Au NP particles with a size range close to the minimum of the threshold laser fluence are most difficult to obtain from laser irradiation. Although this size fraction is often obtained as a second mode after PLAL (depending on laser parameters) [Kabashin2003], this particle size regime can hardly be synthesized by further laser irradiation of the colloid, given that particles in this size regime will interact at the lowest laser fluence threshold and thus be first transformed into different sizes.

Therefore, accessing gold nanoparticles with 50 nm in diameter that have the advantage of high purity as known for laser-generated particles will be challenging. One possibility is to separate the 50 nm particles, formed by PLAL without size quenching by fractional centrifugation [Bonaccorso2013]. On the other hand, Au NP in the size regime of 50 nm are a perfect educt for fundamental PLFL studies, as shown by Hashimoto et al. [Hashimoto2012].

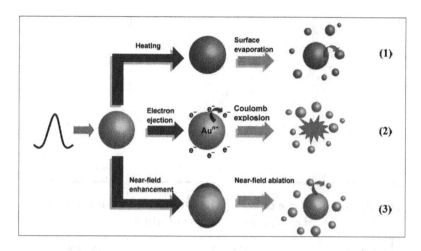

Figure 7: Mechanisms for fragmentation of gold nanoparticles excited by pulsed lasers taken with permission from [Hashimoto2012]

Werner et al. showed that monodisperse gold nanoparticles can be generated with high reproducibility using high pressure chambers for laser irradiation of 100 nm gold and silver nanoparticles [Werner2013]. Hashimoto and co-workers proposed a general fragmentation model for the interaction of laser light with plasmonic particles, as shown in Figure 7. They distinguished three possible mechanisms for change of plasmonic particles size. One is release of small particles by vaporization of particles' surface due to sufficient heating of the particle. Another model is the coulomb explosion, occurring when several electrons are effectively dislocated from their atomic partner in a small area. This results in the atoms' strong repulsion to each other and causes disruption of the atomic structure, the so-called coulomb explosion. As proposed, the effective excitation of the surface plasmons can also result in a release of nanoparticles. This near-field ablation mechanism demands laser wavelengths close to the surface plasmons' wavelength. The model has recently been refined by Strasser et al., correlating particle temperature with the change in size and shape [Strasser2014]. Nonetheless, independent of the occurring mechanism, if particles are fragmentized they will undergo subsequent ripening and thus particle sizes will increase after fragmentation. To avoid this, one possibility is size quenching by organic ligands, although this will cover the particle surface, which might hinder later applications. Hence, investigations are required to address how the size limit of around 4 nm can be overcome, thus fabricating ultra-small and ligand-free gold nanoparticles.

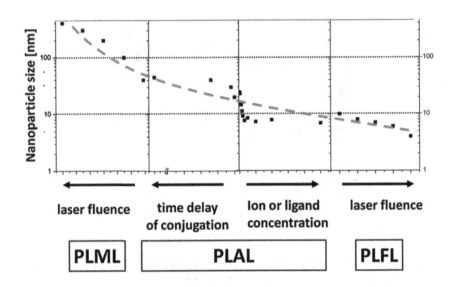

Figure 8: Achievable sizes for laser-generated gold nanoparticles by different strategies (taken with permission from [Asahi2015]) adapted from [Rehbock2014]

These particles hold particular interest as below 5 nm particle diameter relevant catalytic activity starts [Haruta1997] and the origin of fluorescence on gold atom clusters (<3 nm) remains under debate [Goldys2012].

A recent review from Rehbock et al. summarizes the possibilities of size control for pure metal and metal alloy nanoparticles using pulsed laser irradiation [Rehbock2014]. This review demonstrates that a wide size range from a few hundreds of nanometers down to around 4 nm can be addressed by treatment with pulsed lasers.

2.6 Fundamentals of pulsed laser melting in liquids

Pulsed laser melting in liquids (PLML) is a technique to generate spherical particles in the sub-micrometer range by laser irradiation. In contrast to laser fragmentation where educt particle size is reduced, laser melting causes particle fusion and reshaping to sub-micrometer spheres. First reported and established by Naoto Koshizaki and co-workers [Wang2010, Ishikawa2007], this technique is gaining increasingly attention, becoming an emerging research field. PLML enables changing particles' structure (such as size and shape) and the chemical composition, both of which depend on the applied laser fluence. To provide a comprehensive overview of the current state of the art, this chapter is divided into the influence of PLML on particle chemistry and the influence of PLML on particle size and shape.

Influence of PLML on particle chemistry
Following their pioneering studies Wang et al. investigated different materials for PLML, such as zinc oxide sub-micrometer spheres (SMS) [Wang2011], and Ishikawa et al. reported about boron carbide SMS by laser irradiation of boron particles in ethyl acetate [Ishikawa2007]. For this, they assumed a mechanism [Ishikawa2010] proposing that in the region of the focal point the boron is transferred into boron carbide and in lower fluence regimes the boron carbide particles fusion to SMS. They suggest a laser-induced development of a carbon layer onto educts' (boron nanoparticles) surface prior to a followed laser melting. During their experiments, they focused the laser beam into the colloidal suspension and differentiated between two regimes. Close to the focal point, the carbonization of boron nanoparticles is assumed and in the regions with lower laser fluences melting is proposed.

Detailed experimental studies from Wang et al. showed that CuO nanoparticles can be partially chemically reduced to metallic SMS. The starting of PLML for CuO was found to be at laser fluence thresholds of around 33 mJ/cm^2 (10 ns, 355 nm) [Wang2012]. They found laser fluences between 17 mJ/cm^2 and 33 mJ/cm^2 as onset for PLML. Above the melting onset fluence and below ~80 mJ/cm^2, no elemental Cu could be observed, but a gradual change of the elemental composition from CuO to Cu$_2$O by increasing fluence. Further increasing the laser fluence, above ~80 mJ/cm^2 results in the formation of elemental Cu with a composition ratio up ~40% of elemental Cu for 150 mJ/cm^2.

In 2013, Swiatkowska-Warkocka et al. reported about bimetallic SMS by fusion of gold and copper oxide nanoparticles [Swiatkowska-Warkocka2013]. They showed that pulsed laser melting in ethanol of Au-NP and copper oxide nanoparticles can generate bimetallic and crystalline SMS. Remarkably, they could synthesize alloy SMS with a composition that shows phase separation in phase diagram.

Another interesting approach was introduced by Li et al. [Li2011a], who reduced Ag_2O to Ag SMS by PLML.

Tsuji et al. recently reported about the influence of surface chemistry for PLML of Au to form SMS [Tsuji2015]. In case of sufficient particle stabilization, the fusion is hindered due to a controlled repulsive surface charge.

Influence of PLML on particle size and shape

Wang et al. found besides the gradual reduction from CuO to Cu that for increasing laser fluence the particle diameter of CuO SMS (or the partially reduced CuO SMS) gradually increases [Wang2012]. Up to 150 mJ/cm², they received smooth SMS particle surfaces, whereas for laser fluences at 150 mJ/cm² particles appear to be rough, indicating that particle fragmentation starts [Rehbock2014a].

Wang et al. also investigated PLML onset for zinc oxide particles, finding it to be between 33 mJ/cm² and 67 mJ/cm², which is also in the regime for our findings [Wang2011]. Later investigations by Wang et al. focused on silver nanoparticles that were transformed into SMS by PLML [Wang2013]. Tsuji et al. investigated PLML of Au NP for laser fluences from 40 mJ/cm² and 100 mJ/cm² [Tsuji2013]. Likewise, Wang et al. found for copper SMS that the size of Au SMS increases with increasing laser fluence. Beside this, Tsuji et al. found that salt-induced educt particle agglomeration is preferential for sufficient PLML of Au NP with 532 nm laser wavelength.

Ceramic materials such as zirconia SMS or hollowed titanium dioxide SMS are reported by Li et al. [Li2012] and Wang et al. [Wang2011a], respectively. Additionally Li et al. reported about silicon SMS [Li2011b].

The current state of PLML shows that particles' chemistry and size are influenced by the laser fluence applied during PLML. Thus, PLML is not only particle melting, fusion and re-solidification as spheres, but it can be. This part of the mechanistic understanding is illustrated in Figure 9, taken from Higashi et al. [Higashi2013].

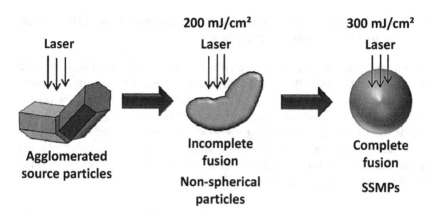

Figure 9: Proposed mechanism for PLML of zinc oxide particles [Higashi2013]

Although exact formation mechanisms are not yet fully understood, early reports on possible applications are available. One possible application has been shown by Fujiwara et al., who used ZnO SMS generated by PLML as a light-emitting laser source [Fujiwara2013]. They added polystyrene particles (900 nm with green fluorescence properties) to a ZnO SMS (synthesized by PLML) film as point defects and showed that the emission threshold for lasing decreases. They describe the lasing effects being improved by PLML of ZnO.

Another recently demonstrated application is the use of SMS as additive for lubricant oils to reduce friction. Hu et al. demonstrated the impact of ZnO, TiO_2 and CuO SMS on reduction of friction coefficients in lubricant oils [Hu2012]. They showed that the friction coefficient decreases if SMS are added to oil, whereas the presence of ZnO educt microparticles in the oil which were not transferred into SMS by PLML showed an increase in the friction coefficient. Liu et al. recently reported about monodisperse Au SMS applied sufficiently for an improved surface enhanced Raman scattering [Liu2015].

To summarize the current state of the art for PLML for sub-micrometer spheres formation, it can be stated that a comprehensive mechanistic understanding of SMS formation is lacking.

Especially the possibility to generate hybrid material compositions and the change in particles' chemical composition holds particular interest and should allow studying the mechanisms responsible for SMS formation. For this purpose, it will be necessary to correlate the applied laser fluence for PLML with the change of particles' elemental composition.

In contrast to laser fragmentation in liquids, detailed studies for PLML thresholds exist, owing to the lower fluences demanded to cause sufficient particle melting. Therefore, unfocused laser beams can be used. In case of focusing, the precise control of applied fluence becomes difficult, possibly reflecting the reason for a lack of sufficient data determining the PLML regime at higher laser fluences. Here, the experimental design developed within this work might come into play to address precisely the high laser fluence regimes, e.g. for precise control of chemistry and melting conditions.

3 Materials and methods

3.1 Materials and instruments

During the experimental work, different lasers, particle materials and analytical instruments were used. This chapter summarizes which particle educts, laser systems and analytical instruments were used within the experimental work, as well as providing an overview of their corresponding specifications. Table 1 lists the powder materials used for PLPPL.

Table 1: Particle materials used for experiments

Material	Supplier	Formular
Zinc Oxide	Sigma-Aldrich	ZnO
Boron Carbide	HC Starck	B_4C
Gold	Allgemeine	Au
Aluminum	Sigma-Aldrich	Al
Copper Nitride	Alfa Aesar	Cu_3N
Copper(I) Iodide	Riedel-de Haën	CuI
Copper(I) Oxide	Riedel-de Haën	Cu_2O
Copper(II) Oxide	Strem Chemicals	CuO
Graphene Nano Sheets	Customized (see 4.4.3)	GNS
Gold Target	Goodfellow (99.9% purity)	

Table 2 lists the apparatus and instruments used within a wide range of experiments. The particular analytic instruments used are also further described in the corresponding research articles cumulated in chapter 4 of this thesis.

Table 2: Apparatus and analytical instruments used for experiments

Instrument	Manufacturer	Use	Model	Description and Parameters
Laser	Ekspla	Laser fragmentation and melting	Atlantic Series	Nd:YAG laser emitting 15 W at 1064 nm and 7.5 W at 532 nm, operating from 100 kHz to 500 kHz, 10 ps
Laser	Coherent	Laser fragmentation and melting	AVIA	Nd:YAG laser emitting 23 W at 355 nm with 85 kHz, 40ns
Analytical Disc Centrifuge	CPS Instruments	Determination of particle size distribution	DC 24000 and UHR 24000	Particle sizing technique in an artificial gravitation field, suitable for a wide particle size range, rotation of the disc from 600 rpm to 24,000 rpm possible, detector wavelength 405 nm or 470 nm
UV-vis Spectrometer	Thermo Scientific	Measuring optical extinction spectrum	Evolution 201	Determination of the extinction from suspended particles in a wavelength range from 1100 nm to 190 nm with 1 nm bandwidth
Power Meter	Coherent	Measuring laser energy output	Field Max II TOP	
Passage Reactor	Customized Design	Generation of a thin liquid jet for defined particle irradiation	Optimized model (see Figure 12 right)	Liquid jet is generated by a capillary with 1.3 mm inner diameter, liquid velocity approx. 0.6 to 0.3 m/s
Scanning Electron Microscope (SEM)	FEI	Characterization of particle size and morphology	Quanta 400 ESEM	

Instrument	Manufacturer	Use	Model	Description and Parameters
X-ray diffraction (XRD)	Bruker	Crystallographic analysis of particles	D8 Advance	Operating with Cu Kα irradiation
Sonotrode	Sonoplus	Dispersing particles with ultra-sonication	HD2200	Maximum power 200 watt
UV-vis spectrometer	Varian Inc.	Spectrographic analysis of the colloids	Cray 50	Determination of the extinction from suspended particles in a wavelength range from 1100 nm to 190 nm with 1 nm bandwidth
Transmission electron microscope (TEM)	JEOL	Characterization of micro/nano hybrid structures	JEM-2100	Operating with high resolution at NIMS, Tsukuba, Japan
Transmission electron microscope (TEM)	Zeiss	Characterization of micro/nano hybrid structures	EM190	
Dynamic light scattering (DLS)	Malvern	Determination of particle size distribution	Zetasizer Nano ZS	
Laser	SPI Lasers PLC	Laser sintering	G3 red ENERGY, SM-series	Operating with continuous wave
Focused ion beam (FIB)	FEI	Cutting a lamella from the sintered hybrid material	Helios NanoLab 600, DualBeam FIB/SEM	
Gas chromatograph (GC)	Shimadzu	Characterization of the gas composition	GC-2010plus	Operating with argon as reference gas

For the fabrication of gold nanoparticles adsorbed on the zinc oxide microparticle support different ratios of colloidal gold nanoparticles to zinc oxide microparticles were brought into contact. The amount of gold nanoparticles for compounding of zinc oxide and gold is given in weight percent (wt%) related to the mass of zinc oxide particles. Thus, the amount of gold referred to the compound material (ZnO+Au) is lower. The following diagram shows deviation of the weight percentage determined with reference to added support material and the absolute weight percentage of the compound formed from nanoparticles and support. This is negligible for loadings up to 10 wt% but becomes significant for high loadings (above 20 wt%).

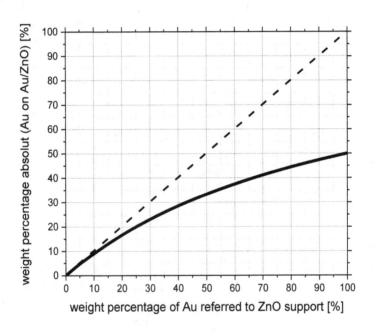

Figure 10: Deviation of weight percent referring support material (ZnO) to absolute mass percentage of the compound (Au/ZnO)

For laser irradiation, the applied laser fluence holds particular interest. This describes the energy density brought to the particles' surface and can be determined by dividing lasers pulse energy with the spot area. To determine the laser fluence in the focal point, the minimal possible laser spot has to be calculated. Given that is not possible to focus laser light to an infinite small area, a correlation between the laser wavelength, focal length of the lens, beam diameter before focusing and beam quality is given in an equation. The focal spot diameter can be calculated by:

$$d_f = \frac{4 \cdot \lambda}{\pi} \cdot \frac{f}{D} \cdot \frac{1}{K}$$

With: d_f: focal spot diameter, λ: laser wavelength, f: focal length, D diameter of raw laser beam before focusing

Where for K, the beam quality number, the following correlation holds:

$$K = \frac{1}{M^2}$$

With: M^2 (beam parameter product): ratio of beam parameter product to an ideal Gaussian mode where a value of 1 describes an ideal Gaussian beam

All lasers used for the experimental work operate in a TEM_{00} mode. Figure 11 illustrates the beam profile emitted from the used ps laser under experimental settings. This shows that the emitted beam profile is close to an ideal Gaussian profile. To compare the laser intensities caused by application of different laser pulse lengths, the laser power per pulse and area might be considered; thus, 30 J/cm² for 10 ps result in 3 TW/cm² and 600 J/cm² for 40 ns in 15 MW/cm². This explains why different pulse lengths result in different laser melting and fragmentation thresholds regarding the laser fluence.

Hence, mechanisms and processes occurring for nano- and picosecond PLPPL will differ.

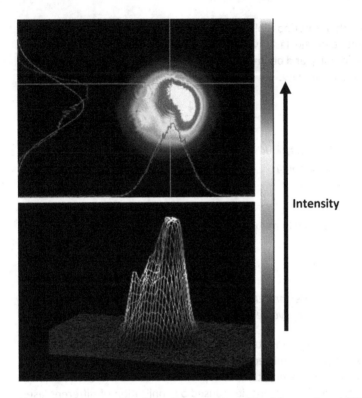

Intensity

Figure 11: Beam profile of the picosecond laser (Ekspla) operating at 532 nm with 100% power and 100 kHz (settings for the PLFL experiments)

3.2 Experimental design

A fundamental aspect of this work is the experimental design of the fluid handling reactor, affecting both the laser fluence gradients and the defined volume flow. As discussed in chapters 2.5 and 2.6, mechanisms that occur during laser fragmentation or laser melting strongly depend on the applied laser fluence. Therefore, it is necessary to control the laser fluence as precisely as possible. The first design of a fee liquid jet used for laser fragmentation was introduced by Wagener et al. [Wagener2010a]. This first design comprised a nozzle that was fed with the suspension by a pump and connected with tubes. The liquid that passed the nozzle was collected in a bottle and recirculated into the pump. Unfortunately, this design combines the design of a flow reactor (part of the liquid jet) with a stirred tank reactor (the part where the suspension is collected and taken by the pump for circulation). Thus, the particle properties achieved during PLPPL additionally depend on the volume and processing time, besides diverse parameters such as educt particle size, particle concentration, optical properties of particles and laser parameters.

To simplify the idea of a free liquid jet, a glass reactor that allows sequential irradiation was designed. This design enables the suspension to pass the illumination zone and thus it can be characterized stepwise. Furthermore, no pump or tubes are necessary, which is often preferential for colloidal stability as the exposure of the colloid to different materials that can potentially release ions (which can result in colloidal instability) is reduced to a minimum. Pumping sometimes also results in colloidal instability, especially for colloids which particle stability is based upon electrostatic repulsion, only. For this design the thinnest capillary diameter, giving a continuous flow without external pressure, was determined. A diameter of 1.3 mm was experimentally found to be appropriate for the formation of a continuous flow transporting particle dispersions with educt particles up to several micrometers. Kuzmin et al. used the idea of this design for their experimental studies [Kuzmin2014] with the capillary outlet at the side of the reservoir. However, as preliminary experiments showed that no significant sedimentation occurs within the reservoir´s residence time for a single passage (~30 seconds for 50 mL) for the used micro- and sub-microparticles used a further improved design was used for the experiments, where the capillary is in the middle bottom of the reservoir. This simplified design allows a complete drain of the reactor and was used for most of the experiments. Figure 12 depicts illustrations of the conventionally-used setup for PLFL and PLML (a) and the setup developed within this research work (b). In Figure 12 c), a schematic magnification of the laser irradiation zone is shown with the process parameters required for additional characterization.

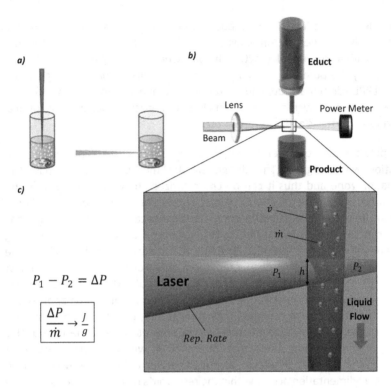

Figure 12: Illustrations of the a) conventionally-used experimental designs for PLFL and PLML (taken with permission from [Sylvestre2011]), b) passage reactor design and c) magnification of laser irradiation area with parameters required for process characterization

It can be stated that a simple design reducing the process parameters that have an influence on PLPPL and on the occurring mechanisms is preferential. With this design a defined laser fluence can be applied what is required to fundamentally study laser fragmentation and melting of suspended particles. The passage reactor enables energy balancing giving the specific energy input by measuring the power loss created by the particles, as illustrated in Figure 12. For most of the experiments, a volume of 50 mL particle suspension with 0.1 wt% particles was used. Filling the reactor with 50 mL resulted in a liquid jet velocity of 0.35 m/s (50mL) to 0.325 m/s (end of draining) using the capillary diameter of 1.3 mm. A diagram depicting the velocity in dependence of the volume is shown in Figure 8-51.

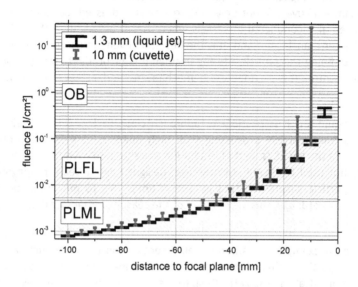

Figure 13: Fluence deviation during beam propagation through a 1.3 mm (liquid jet) and 10 mm (standard cuvette) liquid layer for a 100 mm lens, 3.5mm raw beam diameter for a laser source with 532nm, 10 ps and 15 Watt

Depending on the focal length of the lens used for focusing the laser beam, the laser parameters and path length of focused beam, a particular propagation through the particle suspension occurs, resulting in a fluence regime in the liquid for particle irradiation. The resulting fluence regime for the passage reactor (with a 1.3 mm capillary forming the liquid jet) compared to a standard cuvette irradiated from the side (with a propagation length of 10 mm) is illustrated in Figure 13. Fluences were calculated for the laser parameters used to study PLFL of ZnO. In the diagram, different regimes such as PLML, PLFL and the optical breakdown (OB) of the liquid are marked. Detailed experimental results for determination of these regimes are given in chapter 4.1.1. The mechanisms occurring during laser irradiation of suspended particles strongly depend on the laser fluence. Therefore, it is necessary to minimize fluence deviations within the irradiated volume.

It is obvious that especially for high fluences in case of the 10 mm beam path length (standard cuvette), not only PLFL regime is addressed, but also the OB. PLFL is often also conducted in vessels where beam propagation through the liquid is longer than 10 mm and thus the deviation even increases. A thin liquid filament confining the particles during irradiated is favorable to study process efficiency, control the fragmentation mechanisms and achieve reproducible results.

With the liquid jet, it is possible to approach close to this OB regime, thus enabling defined particle processing with high fluences to avoid the optical breakdown of the liquid. Note that irradiation of a standard cuvette from its side is usually not used for PLPPL. Mostly used designs have a beam propagation length in particle suspension that is several times the 10 mm used for this calculation. Accordingly, this may be one of the reasons for lacking data of PLFL thresholds.

To provide an example, Nakamura at al. recently reported photoluminescent K_2SiF_6:Mn^{4+} particles that showed enhanced luminescence properties after fragmentation [Nakamura2014]. Despite using a cuvette as a vessel for fragmentation, they note in their manuscript the idea and the set-up shown in Figure 12 b), namely the irradiation in a free liquid jet. They considered fluence deviation that occurs during beam propagation through the cuvette. In their study, they showed that deviation from highest to lowest laser fluence is around 20%.

4 Results and discussion

This chapter is divided into three sections and a further ten sub-sections, addressing different experimental results and theoretical aspects of laser fragmentation. 4.1 addresses the findings for laser fragmentation, 4.2 the laser melting process and 4.3 derives key figures from the experimental investigations.

First, the influence of laser fluence on PLFL and specific energy input is investigated and studied in chapter 4.1.1. Here, a key parameter to evaluate process efficiency is derived from optical extinction spectra. Accordingly, different regimes and mechanisms occurring during laser fragmentation can be observed, distinguished and correlated to specific laser fluences. These investigations are mainly conducted on zinc oxide microparticles as model substance characterizing PLFL, including the specific energy input, in the designed passage reactor. Further optimized fragmentation conditions are transferred to boron carbide microparticles, where interesting effects regarding particles morphology can be observed, leading to the conclusion of different mechanisms underlying the formation process.

These investigations follow a determination of process limitations in chapter 4.1.2 regarding achievable particle size and the influence of particle concentration on laser fragmentation. The observations and findings show that fragmentation efficiency and nanoparticle yield underlies optimization, where higher educt particle concentrations result in higher yield yet also cause an attenuation of the PLFL process. The possibility to amplify laser fragmentation is discussed in the subsequent chapter (4.1.3), where gold nanoparticles are used as plasmonic antennas.

In the next section (4.1.4), laser fragmentation of gold nanoparticles in the presence of an oxidative reagent is investigated, whereby ultra-small gold nanoparticles (gold atom clusters) could be synthesized. The redox potential and amount of the oxidative reagent led to the conclusion that sufficient gold particle stabilization could be achieved and thus gold nanoparticles could be transferred into ligand-free gold atom clusters. These clusters were exemplarily transferred to support relevant for catalysis, showing that they can serve as building blocks for heterogeneous catalysts. The impact of an oxidative reagent indicates that surface chemistry is essential for the particle size and colloidal stability obtained.

Additionally, it is shown that a change in chemical composition of particles can be achieved, as addressed in section 4.1.5, where results of different copper compounds and obtained nanoparticles that were fragmentized are shown. Here, particles are characterized regarding their chemical composition, providing a significant insight into possibilities of pulsed laser fragmentation. It is indicated that mechanisms occurring strongly depend on optical particle properties, which determine the absorption efficiency of microparticles.

While the experimental results for the different copper compounds indicate that laser-based reduction is possible, they prompt the question concerning whether oxidation might be induced in the same way. The results from 4.1.4 already indicate that oxidation plays an important role for particle stability. Particles can undergo oxidation releasing hydrogen from water after laser excitation of aluminum particles as shown in chapter 4.1.5.2.

Laser fragmentation of particles and laser-induced change of particle composition is only one possibility to change particle properties by PLPPL. Another method is the PLML where particles are exposed to moderate laser fluence values; thus, they are reshaped to form solid spheres with comparable large diameters in the sub-micrometer range. This approach is described and characterized using the free liquid jet technique in 4.7. Here, it was possible to precisely determine fluence thresholds for PLML with 355 nm laser wavelength for the first time. As known from literature, a laser wavelength of 355 nm causes effective melting to generate SMS. However, due to experimental designs, correlation between fluence and mechanism could only be given for metallic particles, to date. For metallic particles, raw laser beams are sufficiently intense to cause fragmentation, although if the beam is focused, no precise differentiation is possible when the path length of the beam through suspensions is not reduced to a minimum. This is introduced in section 4.1 and holds for both PLFL and PLML. Next to the characterization of laser melting, it is shown how hybrid material can be fabricated when micro/nano compounds are molten, using gold nanoparticles on zinc oxide microparticles as in chapter 4.3. This demonstrated how nanoparticles can be embedded into a sub-micrometer sphere and provide access to a variety of hybrid particles, as it should be easily transferrable to other material composition. The last experimental section addresses gold nanoparticles attached to zinc oxide microparticles (similar to

4.1.3 and 4.2.2). In 4.2.3, the influence of gold nanoparticles on additive manufacturing of microparticles (zinc oxide) is shown. Beside the impact of Au NP on the laser processing window, a micro/nano compound is generated embedding gold into zinc oxide with a dispersed ultra-structure.

Finally, key figures are derived from the experiences and experimental results gained with the passage reactor in chapter 4.3. The key figures are described and characterized, demonstrating what should be considered for future up-scaling and process characterization. These process-specific parameters are supplemented with fluence regimes, providing the opportunity for a comprehensive correlation of laser parameters to process conditions.

4.1 Laser fragmentation

4.1.1 Quantification of mass-specific laser energy input converted into particle properties during picosecond pulsed laser fragmentation of zinc oxide and boron carbide in liquids

Synopsis

A detailed study of precise laser fluence and energy balancing is missing in literature due to difficulties in experimental design, as described in chapter 3.2. In this chapter, the determination of thresholds and specific energy input will be studied.

Here, zinc oxide is used as model substrate to study PLFL using the experimental design shown in Figures 12 and 14. By varying the laser fluence applied to the liquid jets surface generated in the passage reactor, different thresholds could be determined by applying precise laser fluences causing only one effect such as optical breakdown of the liquid, particle fragmentation or particle melting.

This enhanced experimental set-up allows quantifying the specific energy input and the correlation of the energy dose with particle properties as size distribution and bandgap energy.

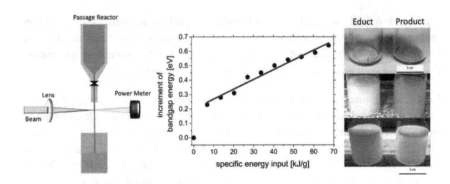

Figure 14: Illustration of the experimental set-up (left), change of the bandgap in dependence of the energy dose (middle) and difference of optical properties from the educt to product zinc oxide particles (right: dried particles, suspended and exposed to daylight, suspended and exposed to UV light (from top to bottom)) for PLFL with a fluence of 0.03 J/cm² and 532 nm [Lau2014a]

During PLFL of ZnO, defects are generated that might hold interest for catalysis [Wöll2007], especially when Cu or CuO particles supported on ZnO are exposed to PLFL, as such compounds are known to be appropriate for catalysis [Behrens2015].

Although laser fragmentation can be transferred to other materials, the observed effects caused by the same laser parameters seem to differ, given that laser irradiation of boron carbide results in mainly spherical nanoparticles generated from the microparticles, whereas ZnO particles underlie a gradual size reduction of the complete particles. Besides the observations and findings regarding the mechanisms during PLFL, it is demonstrated how the developed passage reactor can be easily implemented for laser fragmentation in general and how laser fluences determine the process efficiency.

M. Lau, S. Barcikowski, *Quantification of mass-specific laser energy input converted into particle properties during picosecond pulsed laser fragmentation of zinc oxide and boron carbide in liquids*, Appl. Surf. Sci. 348 **(2015)** 22-29

Abstract

Pulsed laser fragmentation in liquids is an effective method to fabricate organic, metal or semiconductor nanoparticles by ablation of suspended particles. However, modelling and up-scaling of this process lacks quantification on the laser energy required for a specific product property like particle diameter of the colloid or bandgap energy of the fabricated nanoparticles. A novel set up for defined laser energy dose in a free liquid jet enables mass-specific energy balancing and exact threshold determination for pulsed laser fragmentation. By this technique laser energy and material responses can be precisely correlated. Linear decrease of the particle diameter and linear increase of the bandgap energy with mass-specific laser energy input has been observed for the examples of ZnO and B_4C particles. Trends are analyzed by density gradient centrifugation, electron microscopy, UV-vis and X-ray diffraction analysis of the crystal structure. The study contributes to quantitative model parameters for up-scaling and provides insight into the mechanisms occurring when suspended particles are irradiated with pulsed laser sources.

Introduction

In the last two decades after nanoparticle formation by pulsed laser ablation in liquids (PLAL) was reported by Fojtik et al. [Fojtik1993] and Neddersen et al. [Neddersen1993], many efforts to unravel the nanoparticle formation mechanisms were made. Recent results showed that nanoparticles are present already inside the cavitation bubble created by the plasma induced by the laser pulse [Ibrahimkutty2013, Wagener2013]. Further it is known that particle ripening and changes of the laser-generated nanoparticle colloids can occur within miliseconds [Sajti2011] and even minutes up to days after fabrication. Recent publications indicate that significant amount of initially generated species dispersed in the free liquid are not only nanoparticles but even smaller atom clusters [Lau2014b, Lau2014c, Lau2014d]. For PLAL usually a flat surface of the bulk target material is irradiated. Instead of 2-dimensional geometry, PLAL from 1-dimensional thin wire shaped targets releases nanoparticles at the tip of the wire with the cavitation bubble not being confined by a flat surface

[DeGiacomo2013]. This results in significantly higher production rates compared to flat target geometries [Messina2013]. To further enhance the degree of freedom for cavitation and nanoparticle release, 0-dimensional targets made of suspended microparticles can be irradiated and desaggregated or fragmentized into smaller particles. For this pulsed laser fragmentation (PLFL) technique usually a pulsed laser beam is focused into a suspension with micro- and/or nanoparticles to reach the necessary fluences and intensities for sufficient fragmentation. Earlier and recent investigations were focusing on metallic particles [Kawasaki2006, Singh2011, Badr2007, Besner2006, Mafuné2002, Kapoor2000, Amendola2007], whereby studies are more and more including PLFL of alloy particles [Chubilleau2011, Yamamoto2011], semiconductors and organic particulate materials [Zeng2011, Balandin2013, Takeda2014, Sylvestre2011, Nichols2006, Usui2006].

This particle comminution method of pulsed laser fragmentation in liquids (PLFL) could potentially further enhance nanoparticle productivity [Balandin201320, Wagener2012], but modelling and mass balance becomes more complex compared to PLAL, because of the fabricated product mixing with the educt particles. Even though PLFL already was included in the pioneering study of Henglein and Fojtik [Fojtik1993], there is only little information about mechanisms that take place when such particles are irradiated and nanoparticles formed. Shock-wave/mechanical eruption or disruption of the atoms from inside of the particle are suggested mechanisms [Cai1998, Zhigilei1998]. As may be imagined there will be a critical particle size that can be transformed completely into nanoparticles in dependence of the applied beam diameter, laser fluence and intensity. Takeda and Mafuné found this diameter to be around 10 nm for defect-rich cerium oxide nanoparticles in case of 10 ns laser pulses [Takeda2014]. For many target materials ablation thresholds are known, but for dispersed particles it becomes difficult to adjust a well-defined fluence as they often have a curved surface and can move freely in the ambient liquid and further laser intensity is partly screened by the suspension.

Wavelength-dependent investigations of interaction mechanism from nano-particles with pulsed laser sources were reviewed by Pyatenko et al. in 2013 [Pyatenko2013], showing that the applied fluence has a crucial impact on laser melting. Werner et al. reported on the influence of fluence and pulse length for PLFL regarding gold nanoparticles [Hashimoto2012].

Unfortunately, because of the fluence gradients along the beam path it is hard to control the fluence by this technique and balance the input of laser energy. Yan et al. reported structures resulting from sintering (melting) and fragmentation occurring by post-irradiation subsequent to material ablation from a solid target [Yan2010] indicating that both, ablation and melting fluence regimes, responsible for forming these structures may occur in the same vessel because of the fluence gradients. One possibility is to irradiate highly concentrated particle suspension [Balandin2013] to reduce the penetration depth of the laser beam into the liquid and thereby minimize the fluence gradient. But this will make long irradiation times necessary to fragment an adequate amount of particles and at the same time drastically increases the particle-particle interactions. Thus it is very difficult to quantify the laser energy input on the educt particles. A method overcoming these difficulties was firstly introduced by Wagener et al. [Wagener2010a]. Here a free liquid jet confining microparticle suspensions is irradiated. This allows defined intensity threshold determination and balancing educt mass specific energy input. Recently Kuzmin et al. [Kuzmin2014] reported on particle size dependence during laser fragmentation of aluminium powder, using the liquid jet technique, yet without quantitative energy balancing or threshold determination. Takeda and Mafuné reported on the correlation of the bandgap energy of cerium oxide nanoparticles with irradiation time during pulsed nanosecond laser fragmentation in liquid [Takeda2014], but without sampling the effective energy dose taken up by the colloid. The experimental set up in a free liquid jet minimizing axial dispersion of educt particles enables to control the product properties. This is done by differential laser power measurement before and behind the jet. Knowing the laser repetition rate and the irradiated volume with the volume flow rate and particle concentration allows to control the number of laser pulses per number or mass of particles. In our experimental set up each volume element is irradiated by 33 to 667 pulses per passage, depending on the focusing conditions. After the last passage the outflow is captured in a liquid vessel quenching the particle growth. By increasing the number of passages before quenching and final colloid analysis, the mass-specific energy dose can be quantified and correlated to particle properties.

This work investigates the thresholds for laser fragmentation and mass-specific laser energy input to gradually change particle properties of crystalline zinc oxide sub-mirco educt particles. As a semiconductor with a direct band gap around 3.3 eV at 300 K and large binding energy (60 meV) [Özgür2005, Janotti2008] zinc oxide is known to have unique optical and electrical properties. In nano-particulate form and when defects within the atomic lattice structure are induced zinc oxide shows interesting photoluminescence properties [Dijken2000a,

Dijken2000b]. Beside these properties zinc oxide is a widely used and easy accessible material [Özgür2005, Janotti2008].

Additionally the laser fragmentation of boron carbide microparticles was addressed under optimized fragmentation conditions to show that the findings for zinc oxide microparticles can easily be transferred to other materials.

We examine how to determine fragmentation thresholds in a free liquid jet, characterized the energy input and correlated this to energy-specific change of particle properties, demonstrated for the particle size and change of bandgap energy.

Experimental Details

For the experiments a purpose-designed reactor was used, as illustrated in Fig. 15. This passage reactor feeds a 1.3 mm capillary at the bottom of a reservoir where the suspension filament is formed into a liquid jet and is irradiated with defined laser fluences. Particle mass concentration was set to 0.1 wt% in 50 ml of water and pre-treated in an ultra-sonication bath prior to laser irradiation. Subsequent to laser irradiation process the zinc oxide suspension was injected directly from the reactor downstream into 0.1 M sodium dodecyl sulphate (SDS) solution to avoid unwanted particle agglomeration, thus capturing the colloidal state after the laser process and defined number of passages. Note that no surfactant (SDS) was present, only for down-streaming, neither in the educt vessel nor during the laser irradiation. Addition took place only in the downstream product reservoir after the final passage number (up to 100) of each experiment. Laser irradiation was carried out with 75 µJ pulse energy at 532 nm and a repetition rate of 100 kHz using a picosecond laser (Ekspla, atlantic series). Fig. 15 displays how the laser fluence was altered by varying the distance of the liquid filament to the lens (100 mm focal length). Energy input to the particles was determined by the difference of transmitted laser energy from pure water filament and particle suspension filament behind the liquid jet, detected by a power meter (Coherent, Fieldmax TOP II). This calculated laser power input was multiplied by the time of each irradiation cycle (passage) and by referring this value to the particle mass of 0.1 wt% in the 50 ml sample the mass specific energy input is balanced. Hence this passage flow reactor allows the defined and incremental increase of cumulated laser energy dose at constant particle mass. High flow rate of up to 0.6 m/s minimizes axial dispersion and each passaged volume element receives the same number of laser pulses.

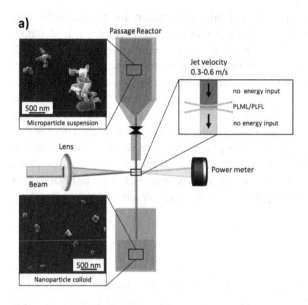

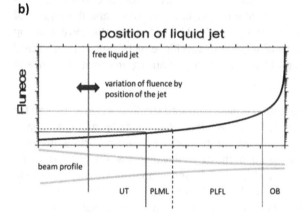

UT: untreated
PLML: pulsed laser melting in liquid
PLFL: pulsed laser fragmentation in liquid
OB: optical break down of liquid

Figure 15: Experimental set up for sequential laser fragmentation in a free liquid jet passage reactor (a) and schematic illustration of the process regimes depending on laser fluence varied by positioning of the jet relative to the focus (b).

UV-vis spectra of the prepared colloids were taken by an evolution 201 spectrometer (Thermo Scientific) in a quartz glass cuvette. Hydrodynamic particle size distributions were detected in an analytical disc centrifuge (CPS Instruments), giving high resolution in a wide particle size range from micro- to nanoparticles. X-ray powder diffraction (XRD) pattern were detected with a Bruker D8 Advance instrument with Cu Kα radiation. Scanning electron microscopy (SEM) pictures were taken with a FEI Quantana 400 ESEM in high vacuum after droplet deposition on carbon thin films. Zinc oxide particles and sodium dodecyl sulphate (SDS) were purchased from Sigma Aldrich and the deionised water was derived from a Millipore installation. Boron carbide (B4C) particles were purchased from HC Starck (Grade HS). For ultrasonication experiments a sonotrode (Sonoplus, HD2200) at 120 W and 20 kHz was used.

Results and Discussion

The results address three regimes occurring for laser irradiation illustrated in Fig. 15: Pulsed laser melting in liquid (PLML) regime for zinc oxide particle suspensions, pulsed laser fragmentation (PLFL) of zinc oxide and the regime of optical breakdown (OB). All these regimes will be defined by the corresponding laser fluence thresholds causing regime-specific effects onto the particles. Experimental emphasis is given on the energy balancing and related change in particle properties.

Finally pulsed laser fragmentation under optimized conditions is investigated for boron carbide particles in order to validate transferability of experimental procedure. From the UV-vis spectra we defined a primary particle index (PPI) [Rehbock2013] of the zinc oxide particles by dividing the relative extinction maximum around 370 nm, specific for ZnO and proportional to its interband absorption, hence to the mass, with the extinction value at 600 nm where big particles, aggregates and agglomerates scatter the light. This PPI value gives good indication whether the particle size is changed to smaller or bigger particles, with high PPI values representing higher fraction of smaller particles. As will be shown in the following, the PPI of the laser-irradiated particle dispersions correlate with the laser energy input, therefore the gradual difference of the PPI of a series of samples treated within the passage reactor filament is determined by the slope of the linear fit from the PPI (mPPI) plotted versus the energy input. Hence, the mPPI is proportional to the effectiveness of the process to fragmentize and desaggregate larger particles/aggregates into smaller primary particles. Fig. 16

and Fig. S 4.1.1 1 (supporting information) summarize the effectiveness of laser fragmentation/desaggregation for the entire process window. The values for the slope of the PPIs were correlated to data from particle characterization via UV-vis spectroscopy, electron microscopy and particle size measurements in the analytical disc centrifuge. PPI values and measured particle properties allowed to differentiate fluence regimes of laser processes (Fig. 16), described in the following in order of increasing laser fluence.

Regime of laser melting - PLML

If the picosecond laser fluence on the surface of the liquid jet exceeds 0.9 mJ/cm² pulsed laser melting in liquid (PLML) occurs, causing fusion and reshaping [Wang2010] of the initially aggregated educt particles. This PLML process naturally results in a decrease of the PPI, giving negative values for the slope m_{PPI}. To our best knowledge this is the first PLML shown for ultrashort laser pulses. Note that we did not operate in the UV wavelength as usual in PLML of semiconductors but at 532 nm. In order to validate if the passage reactor is feasible for the well-established UV-nanosecond PLML [Wang2010, Wang2011], we applied these PLML conditions as well. For this we exemplarily conducted PLML with a 355 nm, 40 ns laser, laser parameters known to fulfil PLML in an effective way [Wang2010]. The corresponding UV-vis spectra and SEM pictures are shown in the supporting information (Fig. 8-2). This confirms that a decrease in PPI values indicates increase in solid particle diameter caused by particle fusion, aggregate melting and re-solidifying spheres. Accordingly, the PPI decreases over the irradiation cycles for UV nanosecond PLML, resulting in negative m_{PPI} values (Fig. 8-2 a).

Typical extinction spectra for the particle responses to 532 nm picosecond PLML with the PPI values as inset are shown in Fig. 17a. With increasing laser energy does the extinction spectra show increased values at larger wavelengths, resulting in decrease of PPI as observed from the nanosecond case. The inset shows that the PPI value linearly decreases with increased deposition of laser energy, equivalent to increase of number of passages. In Fig. 18 SEM pictures of the educt and PLML-treated particles are shown, respectively. Here molten and partially molten particle structures are observed, in agreement with the change of UV-vis spectrum.

Regime of laser fragmentation - PLFL

If the laser fluence is raised above 6 mJ/cm² and kept below 40 - 100 mJ/cm² a significant size reduction of the educt aggregates occurs. Accordingly the primary particle index increases linear with the number of passages or the specific energy input applied for laser fragmentation. Figure 19 depicts the particle size reduction during the process. Confirming the PPI trend derived from the UV-vis-spectra (Fig. 17b), the comminution process via laser fragmentation follows a fully linear behavior as well if the hydrodynamic particle size reduction is plotted versus the specific energy input (Fig. 19a). In the product suspension, two different particle fractions are obtained (Fig. 19b). The bigger fraction is the mass dominant fraction in the product and results from desaggregation of educt particles, exposing the primary particles from the aggregate structure (SEM picture in Fig. 18). The second particle fraction lays around 60 nm hydrodynamic diameter and is fabricated from partial particle ablation mechanism, delivering spherical nanoparticles. From the mass frequency distribution acquired in the analytical disc centrifuge we detected 2 wt% of 60 nm nanoparticles and 98% of desaggregated 180 nm particles from educt for a specific energy input of 67.5 kJ/g.

Due to the sintered morphology of the educt we believe that this is an effective comminution process that may be difficult to achieve by conventional dispersing methods. We compared the PLFL results with strong ultra-sonication with an ultra-sonic finger and results are shown in the supporting information Figure 8-3. PLFL is 5 times more effective regarding the primary particle index and with ultra-sonication is was impossible to decrease the hydrodynamic diameter even at a factor of around 250 times higher mass-specific energy applied for dispersing (PLFL up to 67.5 kJ/g, ultra-sonication: 17,280 kJ/g). From the XRD pattern of laser-fragmented ZnO we calculated the crystallite size (shown in the supporting information Figure 8-4), confirming a crystallite particle diameter reduction from ~250 nm to ~100nm. This supports the mechanistic hypothesis that PLFL is a comminution process breaking down crystals (desaggregation) beyond simple dispersing methods (desagglomeration). Additionally, we determined the change of ZnO bandgap energy from the tauc plots calculated from the UV/Vis-spectra, whereby the square product of the absorption coefficient is plotted versus the photon energy, and the bandgap energy can be determined [Taguchi2009]. The tauc plots are shown in Figure 8-5 and 8-6 in the supporting information. Note that the bandgap energy increase can be correlated linearly with the energy dose (Fig. 19C). Similar behavior has been observed for cerium oxide [Takeda2014] where increased duration of PLFL caused linear change in bandgap, but without balancing laser energy. PLFL allows engineering the bandgap of ZnO precisely by the number of passages in the liquid jet reactor, equivalent to defined mass-specific laser energy input.

The white educt powder is converted into product particles that have an intense yellow color as dry powder (Fig. 19d). Suspended in water, in contrary to the milky ZnO educt particles, the laser fragmented particles show an intense green color if exposed to UV light. Accordingly, Zeng et al. have observed photoluminescence properties arising during PLFL of ZnO as well [Zeng2011]. This photoluminescence indicates high defect density of the fabricated zinc oxide particles [Zeng2010]. An increase in particle defect density is also indicated by enhanced absorption of the particles in the UV range (Fig. 17b). As defect-rich particles are of particular interest for catalysis applications and zinc oxide is known to be an adequate support [Strunk2009], they might be a good candidate for application in heterogeneous catalysis [Haruta1997].

Measured energy input enables balancing and quantification of mass specific energy consumed during the particle comminution, an approach typically used in modelling conventional milling processes. Extrapolation from the experimental data to the total mass of 1g zinc oxide shows that 3 kJ laser energy input would be required to reduce the size by an increment of 10 nm and around 110 kJ for an increment of 1 eV in the bandgap. This holds for the optimized laser parameter and particle mass concentration. Assuming a spherical particle shape and size reduction from 450 nm to 200 nm for the main particle fractions, the created surface by laser fragmentation is around 5 m^2/g zinc oxide. For zinc oxide nanostructures a surface energy of around 1.1 J/m^2 has been reported [Na2010]. Thus the generated surface energy is around 5 J/g by applying 67.5 kJ/g specific laser energy input. As the generated surface energy is around 4 orders of magnitude lower than the optical energy used to create this surface we believe that significant laser energy is consumed by material heating, its chemical conversion (fabrication of defects), and light scattering by the particles and laser-induced cavitation bubbles.

Figure 18 shows SEM pictures of the educt particles and the change in size and shape when PLFL or PLML occurs. The SEM picture for the PLFL process shows educt, product and partially ablated particles present in the suspension at the same time. These two product particle fractions have diameters of 60 nm and 200 nm confirming the two hydrodynamic particle sized observed in the disc centrifuge (Fig. 19).

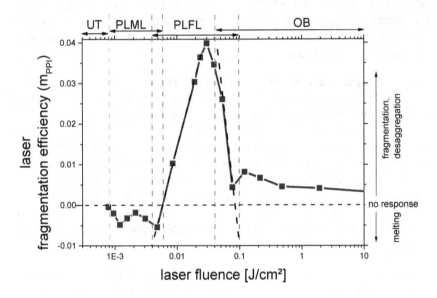

Figure 16: Process efficiency and process windows for picosecond laser fragmentation
of zinc oxide. Diagram shows the dependence of fragmentation efficiency
(change of primary particle index) for the laser fluence applied to the liquid
jet. UT: untreated, PLML: pulsed laser melting in liquid, PLFL: pulsed laser
fragmentation in liquid, OB: optical breakdown of the liquid

The presence of two fractions is of particular interest as this indicates two
different mechanisms forming the product particles. In literature it has been
supposed that (for silicon dioxide particles) short laser pulses induce shock waves
passing the particle and at the same time ablating and vaporizing particles [24].
The shock waves induce desagglomeration and desaggregation resulting in the
larger edged particle fraction. This mechanistic pathway is supported by earlier
data, where pre-milling of ZnO educt particles greatly enhanced PLFL
productivity, possibly by pre-loading stress to the educt, facilitating mechanical
fracturing by PLFL [Wagener2012]. The vaporization pathway may cause the
formation of the observed smaller fraction, the spherical nanoparticles.

Regime of optical breakdown - OB

The fragmentation efficiency decreases drastically when the fluence exceeds
40 mJ/cm² (Fig. 16). At the same time the transmitted energy increases and

fluctuates strongly and vaporization of the liquid is observed. This is accompanied by visible plasma formed at the back of the liquid jet, typical for the optical breakdown of water. At this fluence, the laser energy is transferred into the surrounding liquid medium. For fluences above 100 mJ/cm² the liquid jet is punctuated with a plasma formed by the optical breakdown of water. An influence of the presence of particles on this breakdown threshold was not observed, at least up to a mass concentration of 6 wt% ZnO. Vogel et al. [Vogel1999] observed the laser fluence threshold for optical breakdown in water for 3 picosecond laser with 580 nm wavelength to be at 2.6 J/cm², equivalent to an intensity of about 8 GW/cm². This value is a factor of 40 higher than in our case (50 mJ/cm² for 10 picoseconds pulse length and 532 nm). Deviation can be explained by the curved surface of the liquid jet. The curved surface might contribute to focusing of the laser beam, resulting in a higher fluence inside the liquid jet.

To study if laser fragmentation under optimized conditions is transferrable to other particle materials we investigated the picosecond laser fragmentation of boron carbide microparticles in the same way. The black B4C powder was dispersed in pure water at the same particle concentration like the zinc oxide particles investigated before (0.1 wt%) and was exposed to the PLFL fluence regime under the conditions where the highest value of mPPI (fragmentation effectiveness) was observed for ZnO (30 mJ/cm²). Again, linear increase of primary particle index with increasing picosecond laser energy input was observed, indicating a well-controlled comminution process. However, in contrary to ZnO, the majority of boron carbide particles are transferred into spherical nanoparticles. These nanoparticles dominantly form large hydrodynamic agglomerates in the analytical disc centrifuge (Fig. 8-7), but SEM pictures confirmed the fabrication of spherical nanoparticles. A particle size histogram from the SEM pictures of the educt particles and the fabricated nanoparticles is shown in Fig. 20. It can be seen that PLFL of boron carbide particles leaves spotty craters and holes with a diameter of around 100 nm on educt particles surface (Fig. 20c). Interestingly, the product particles have same diameters like the craters left on the larger educt particles. This is of particular interest as the laser beam can by no means be focused to this diameter with common focusing optics. We believe that the fabricated product nanoparticles are acting as nano-lenses focusing the laser light in the near-field onto the educt microparticles. Fig. 21 gives an overview of the proposed mechanisms occurring for PLFL.

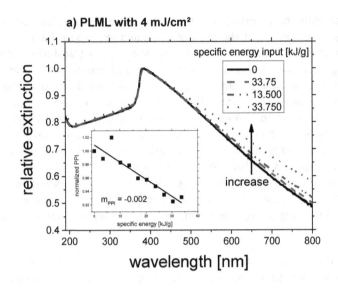

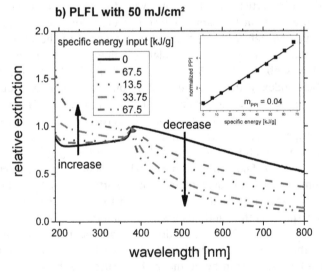

Figure 17: Effect of mass-specific energy input on the extinction spectra of the ZnO
particle suspension (diagrams) and on the primary particle index (inserts).
The extinction values are peak-normalized and the PPI values are normalized
to the educt PPI. Left: laser melting (PLML) regime, right: laser fragmentation
regime (PLFL)

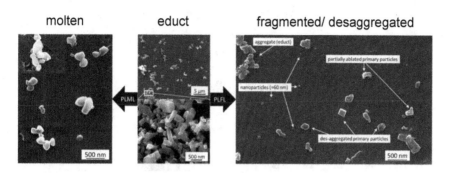

Figure 18: SEM pictures of the zinc oxide educt particles (middle), particles formed by PLML (left, 4 mJ/cm²; 33.75 kJ/g) and particles irradiated with 40 mJ/cm² laser fluence and 67.50 kJ/g specific energy input in the PLFL regime(right).

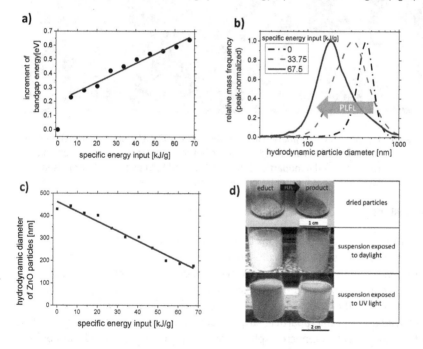

Figure 19: Particle size and bandgap engineering by laser fragmentation of ZnO in liquid flow with controlled energy dose. In a) decrease of hydrodynamic particle diameter is shown. Exemplarily particle mass distribution of educt, product and an intermediate smaple are shown (b). This gives evidence that a high amount of particles is treated in the liquid flow. During the fragmentation process the bandgap energy increases linearly with specific laser energy input (c). A photograph of the dried and suspended particles before and after laser irradiation gives evidence for fabrication of highly defect-rich ZnO particles (d).

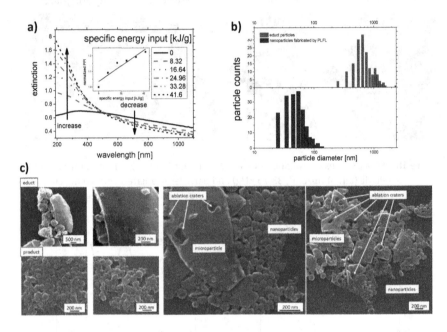

Figure 20: UV-vis spectra, primary particle index (here 500nm/700nm as insert) (a), SEM histogram of boron carbide particles before (top) laser fragmentation and of the fabricated nanoparticles (bottom) (b) and SEM pictures (c) for pulsed laser fragmentation of boron carbide microparticles.

Conclusions

Treatment of particle suspensions by pulsed lasers allows a surfactant-free and wear-free synthesis of submicron spheres and nanoparticle colloids by melting and fragmentation of the educt. However, modelling and scale-up of this laser method requires a robust process design, continuous liquid handling and defined irradiation conditions regarding the fluence regime and energy dose. Sequential laser irradiation of microparticle suspensions in a passage reactor is a versatile method to fabricate submicron spheres or nanoparticles generated from pulsed laser melting or pulsed laser fragmentation, respectively. Accordingly, this liquid jet technique allows precise energy balancing and threshold determination for pulsed laser fragmentation in liquids. Varying the laser fluence applied to the liquid jet enables to distinguish between different material response regimes.

The threshold for pulsed laser melting of zinc oxide particles was determined at 0.9 mJ/cm² for 10 picosecond pulse duration at 532 nm wavelength. This process has till now only been shown for nanosecond pulse durations (in the UV), but for low fluences pulsed laser melting in liquid is possible as well if picosecond pulses are applied. Laser fragmentation starts to occur when the fluence reaches 6 mJ/cm² and is most efficient at 30 mJ/cm² close to the optical breakdown threshold of the liquid. The primary particle index defined from the UV-vis-spectra has been introduced as quantitative measure for the fragmentation efficiency and correlates linear with decrease in the hydrodynamic particle diameter. Both colloidal parameters, primary particle index and hydrodynamic diameter, can be linearly adjusted by the number of passages in the liquid flow reactor.

For pulsed laser fragmentation of zinc oxide particles we received highly defect-rich des-aggregated particles and spherical nanoparticles, forming two particle fractions. This indicates that two mechanisms are involved forming these particles possibly divided into mechanical and vaporization pathway. Beside the change of particle size laser fragmentation allows the bandgap engineering of the semiconductor. Hereby the measured bandgap energy of the colloidal suspension can be controlled by the specific energy input and correlates linearly with the number of passages. After laser energy input of 67.5 kJ per gram zinc oxide a yellow powder is obtained after drying. This indicates highly defect-rich zinc oxide particles which are stable over months under atmospheric conditions and at room temperature. Therefore laser fragmentation can be used as a tool to adjust the bandgap energy of a particulate material. Further pulsed laser fragmentation of boron carbide microparticles under optimized parameters shows that the fragmentation process in a continuous liquid jet can easily be transferred to other particle systems with defined energy dose. The linear behavior of the continuous reactor, correlating aimed particle properties with number of passages is ideal for modelling reactor design for a defined product. But even more important, mass-specific energy data are hereby available allowing to scale the laser power required for a demanded throughput. Taking both together, pulsed laser treatment of educt particles in liquid is a particle fabrication technique, when modelling particle response and up-scaling production parameters are easily accessible. The observed linear behavior (at least of the two investigated materials) is ideal for robust operation of this production method.

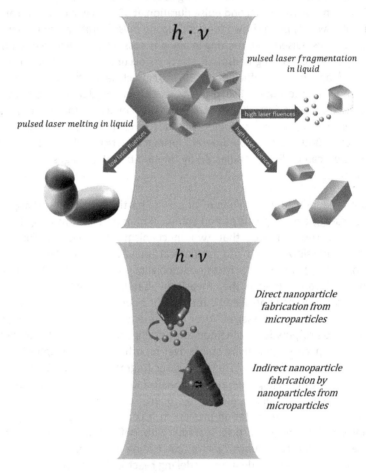

Figure 21: Proposed mechanism for nanoparticle formation from zinc oxide microparticles (top) and from boron carbide particles (bottom). In case of low fluences ZnO microparticles partially melt down and re-solidify. For higher fluences particles are fragmentized and vaporized. For boron carbide mircoparticles small ablation craters are observed on the microparticles that could be created from the nanoparticle interaction with the laser beam.

4.1.2 Influence of particle mass on laser fragmentation efficiency and process limitations

After addressing process efficiency in dependence of laser fluence and specific energy input in the previous chapter, here the influence of particle mass concentration and process limitations are described. Zinc oxide particle suspensions with 0.01, 0.05, 0.1, 0.5, 1, 2, 3, 4, 5, 6 wt% were fragmentized, applying an optimized laser fluence of 0.03 J/cm² on the surface of the liquid jet in the passage reactor. Laser fragmentation efficiency is determined by the primary particle index increase per number of passages (m_{PPI}), similar to section 4.1.1. To obtain this value of the PPI, the extinction maximum around the bandgap energy is divided by the extinction value at 600 nm for each sample taken after different numbers of passages. Figure 22 shows the fragmentation efficiency plotted versus the reciprocal particle mass concentration. For ZnO, particle mass concentrations above 0.5 wt% the laser fragmentation process of microparticles in liquids becomes inefficient. A dose response fit was found to give good correlation of the laser fragmentation efficiency. This can be explained by attenuation of laser light at high concentrations and no further efficiency increase for low particle concentrations. A second ordinate is drawn on the left side in the diagram, showing the relative laser fragmentation efficiency for the investigated particle mass concentrations.

The existence of a maximum laser fragmentation efficiency at fixed experimental parameters for different mass concentrations can be explained with help of the passage reactor design; for instance, assuming that only a single particle is irradiated in the illuminated volume in the passage reactor while the flow through. The process efficiency will not change if in a subsequent volume unit another single particle will be irradiated. Therefore, it will have no impact if one particle is suspended in the liquid or two or more, for instance. However, with increasing particle concentration there will be a change in process efficiency when particles shield each other to the incident laser light.

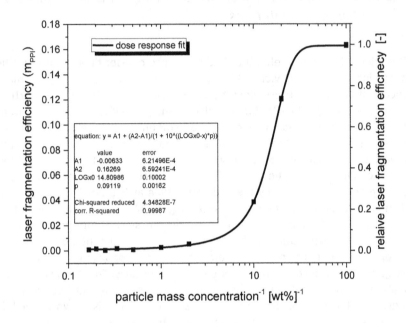

Figure 22: Influence of particle mass concentration on the laser fragmentation
efficiency determined under optimized fragmentation conditions for zinc
oxide microparticles (laser parameters: 10 ps pulse length, 532 nm
wavelength, 0.03 J/cm²) in the free liquid jet with a dose response fit

Thus, the number and volume of particles in the interaction zone where the laser
irradiates them strongly determines the overall process efficiency. Therefore,
this process is subject to an optimization problem between effectivity of
fragmentation (favored by low wt%) and nanoparticle yield. For the experiments
presented in the previous chapter, 0.1 wt% was found to be effective in the
laboratory scale, which already allows detecting the change of particle properties
by UV-vis extinction spectroscopy after one irradiation cycle. Although the
absolute value of laser fragmentation efficiency depends on optical particle
properties such as the extinction coefficient at the laser wavelength used, as well
as on educt particle size, the influence of particle mass concentration on change
of process effectivity might be similar for other non-metallic particulate
materials.

A further enhanced experimental design to study laser fragmentation mechanisms more precisely would be an irradiation of a single particle in the laser-liquid interaction zone by a single laser pulse. In chapter 4.3 where key figures are derived the possibility how to implement the passage reactor designed and used for this purpose is discussed. A first approach is the decrease of particle mass concentration and investigation of particle properties such as UV-vis extinction and size.

Therefore, to determine the achievable particle size for PLFL of zinc oxide microparticles in the passage reactor, a suspension of 0.01 wt% zinc oxide in water was fragmentized for up to 100 passages under the previously-described parameters.

Figure 23 shows the UV-vis spectra for 0 to 30 passages (a), 30 to 100 (b) passages and the resulting primary particle index for different passages (c). After linear increase until 30 passages the PPI decreases after 30 passages. To understand this, the change in particle size has to be considered.

Due to the low particle concentration and low extinction of the small particles (below 10 nm, see Figure 25), the results from the analytical disc centrifuge were not very precise, although they show a gradual size reduction of the main fraction (around 450 nm). The results for disc centrifuge measurements after 0, 5, 10 and 20 passages are shown in the Appendix Figure 8-8. The decrease of the PPI after 30 passages can be explained by fabrication of small and e.g. amorphous nanoparticles that contribute to neither the extinction around 350 nm nor the extinction at 600 nm.

The degradation in particle size becomes significant after 30 passages and leads to a decrease in the PPI. This observation is confirmed by TEM pictures showing the zinc oxide particles after different passages. Figure 24 summarizes the experimental procedure and change of optical properties of the suspension after laser fragmentation. The fragmented suspension appears transparent and TEM images reveal small nanoparticles with diameters below 10 nm. Figure 25 shows typical TEM images after 0, 20, 80 and 100 passages on top and the corresponding UV-vis spectra and histograms derived from several TEM images. Interestingly, the particle size slightly increases from passage 80 to 100.

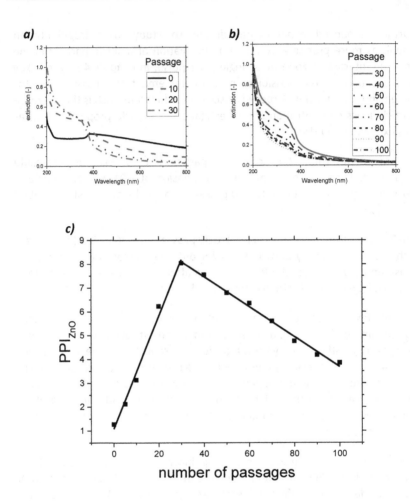

Figure 23: a) UV-vis spectra from 0 to 30 passages; b) UV-vis spectra from 30 to 100 passages; c): primary particle index (PPI) for pulsed laser fragmentation under optimized laser parameters at 0.01 wt% zinc oxide particle mass concentration

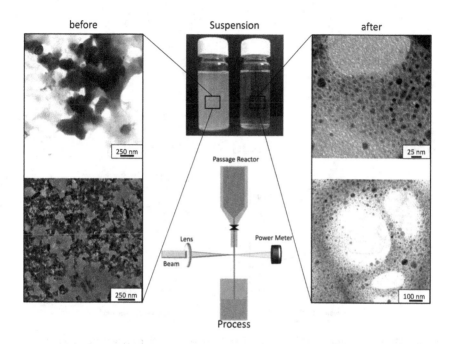

Figure 24: TEM images and photograph of the zinc oxide colloids in water with a
particle mass concentration of 0.01 wt% before (left) and after 100 passages
laser fragmentation (right). The middle bottom shows the experimental set-
up.

One possible explanation for particle size increase can be the ripening due to
colloidal instability, because the size capturing surfactant SDS was added just
after the last passage. Therefore, the particles obtained after 100 passages had a
longer reaction time for aggregation and agglomeration.

Although the reason for slight particle size increase between 80 and 100 passages
cannot clearly be stated here, it is shown that particle concentration during laser
fragmentation has a crucial influence on the process efficiency. Experiments with
low particle concentrations (0.01 wt%) demonstrate that small (around 10 nm)
zinc oxide nanoparticles can be generated from microparticles. Note that the
microparticles quantitatively converted into the nanoparticles. No precipitation
was observed nor were educt-product separation steps executed.

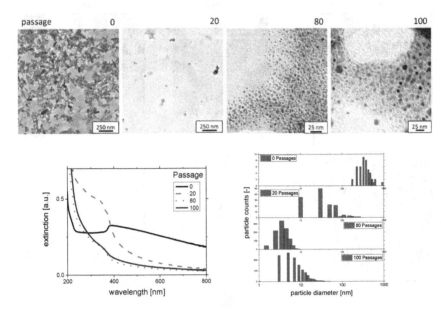

Figure 25:   Representative TEM images after 0, 20, 80 and 100 passages (top), UV-vis
                 spectra of the corresponding suspensions and TEM histograms derived from
                 several TEM images

From the determined solubility (see chapter 2.1), it should be excluded that ZnO
particles dissolved given that only around 18 wt% of the particles are soluble at
a pH of 6.7. One possibility might be that ZnO particles are dissolved by intense
laser irradiation and a super-saturated solution is captured by the addition of
surfactant after the last passage. This would be in agreement with the seed
diameter reported by Schindler et al. [Schindler1965].

This shows that laser fragmentation offers good potential for nanoparticle
generation and that under precise controlled process conditions quantitative
conversion from micro- or sub-microparticles to nanoparticles is also possible. An
educt-product separation is often difficult. Filtration steps mostly result in a low
nanoparticle yield, especially when only electrostatically stabilized nanoparticles
are filtrated. For an educt-product separation of a bimodal size distribution
sedimentation is superior due to a possible cut off size in the gap of particle
modes. In case of a gradual size reduction as observed here for zinc oxide

particles a cut off size that can have an overlap in the particle modes has to be chosen. Thus, quantitative conversion of educt particles at lower concentrations followed by a subsequent concentration increase through vaporization of the liquid phase might be preferential to achieve monomodal and high concentrated nanoparticle suspensions from PLFL. A quantitative conversion leads to a sufficient yield and renders additional processing such as size selection unnecessary.

Investigating and studying the influence of particle mass concentration on PLPPL is crucial for process understanding and potential up-scaling. Another important question arises when the presence of a second particulate material is considered, e.g. plasmonic nanoparticles. Accordingly, this will be discussed in the following.

4.1.3 Laser fragmentation of supported particles: gold on zinc oxide

As described in chapter 2.4 and 2.5, gold nanoparticles absorb laser light around the second-harmonic wavelength of Nd:YAG lasers very efficiently, due to their surface plasmon resonance (SPR). The idea of a gold-amplified laser fragmentation is to use gold nanoparticles attached to microparticles surface as plasmonic antennas. For this purpose, laser-generated gold nanoparticles are adsorbed onto microparticles. These ligand-free Au NP are preferred due to their bare surface and electrostatic potential. Due to their charge, high concentrations of Au NP can be adsorbed with 100% efficiency to particulate materials that have an opposite zeta potential [Marzun2014]. Further details on the preparation method for gold nanoparticles supported on zinc oxide microparticles are given in the experimental section of chapter 4.2.3. Figure 26 shows the influence of pH value on electrostatic potential (zeta potential) of particles surface is shown for gold nanoparticles and zinc oxide microparticles. Under acidic conditions, zinc oxide dissolves [Degen2000]. The isoelectric point (IEP) of ZnO microparticles is around 8.5 [Degen2000], while the IEP of gold nanoparticles is around 2 [Sylvestre2004a], [Zeng2005], [He2008]. To attach the gold nanoparticles to zinc oxide microparticles, their net charges should be opposing, namely zinc oxide positive and gold negative. Therefore, Au NP can be adsorbed to zinc oxide microparticles from pH values of > 2 to < 8.5, reflecting a wide range covering the neutral regime [Marzun2014]. To support the Au NP on ZnO MP, pure water was used as a solvent. Note that zinc oxide should not dissolve, meaning that acidic conditions should be avoided. To provide a large area of zinc oxide hydrodynamically accessible by the Au NP, ultra-sonication should be applied during the supporting procedure. Due to the size of microparticles, the micro/nano hybrid compound sediments in the reaction container leaving pure water as supernatant. This allows the easy removal of nanoparticles from their liquid environment when attached to microparticles' surface. After additional drying, a purple powder is obtained (see 4.8 and [Marzun2015]). To investigate the influence of gold nanoparticles supported to zinc oxide microparticles during pulsed laser fragmentation, these powders were subjected to 532 nm picosecond PLFL at different amounts of gold nanoparticle loads.

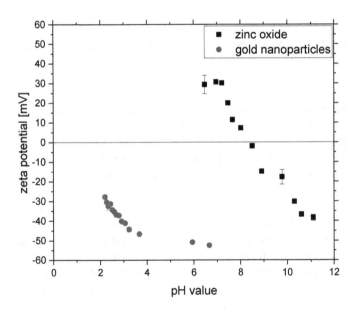

Figure 26: Zeta potential at different pH values of zinc oxide microparticles and gold nanoparticles

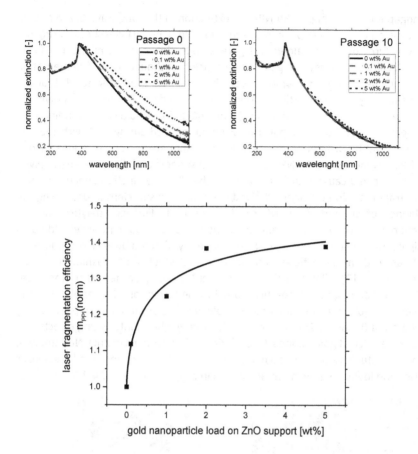

Figure 27: UV-vis extinction spectra of ZnO microparticles with different gold
              nanoparticle load before and after 10 passages (top left and right,
              respectively) and laser fragmentation efficiency for different amount of ZnO-
              supported gold nanoparticles (bottom)

Figure 27 (bottom) shows the change of laser fragmentation efficiency in
dependence of the gold amount supported on zinc oxide. Here, the efficiency was
determined analogous to chapter 4.1.1 (defined as the slope of linear fit from the
primary particle index m_{PPI}). In Figure 27 extinction spectra determined by UV-vis
spectroscopy of the different educt powder materials (gold load on ZnO of 0 wt%,
0.1 wt%, 1 wt%, 2 wt%, 5 wt%) suspended in water before PLFL and after 10
passages is shown. Spectra are normalized to the local maximum around the

bandgap energy. At higher Au NP load extinction in the long wavelength regime is increased due to the optical properties of gold nanoparticles. Considering the SEM pictures of educts shown in Figure 28 left, this increase might be caused from locally agglomerated gold nanoparticles. Although laser fragmentation efficiency derived from m_{PPI} increases, this could be attributed to PLFL of the aggregated gold nanoparticles. This is confirmed in Figure 28, showing a representative SEM image of 5 wt% Au on ZnO before and a corresponding TEM image after PLFL. Gold nanoparticles are fragmentized during PLFL with 532 nm. It cannot yet be distinguished whether gold nanoparticles enhance PLFL of ZnO or if PLFL removes the Au NP from their ZnO support and disperses them, given that both could cause a similar change in the UV-vis extinction spectra. Hence, laser fragmentation efficiency derived from m_{PPI} is insufficient in describing the influence of supported Au NP on ZnO for PLFL. Further analytics such as fractioning of particle sizes and elemental analysis could provide additional insights. However, obviously Au NP are evidently affected by PLFL conditions, as well as ZnO microparticles. Considering the studies of Hashimoto et al. [Hashimoto2012], [Werner2011], [Werner2011a], [Werner2013] for laser fragmentation of gold nanoparticles proving the formation of a nanobubble and the electron-phonon coupling time for gold which is around 6 to 20 ps [Lin2003] and around 0.5 ps for ZnO [Zhukov2012] there might be interesting effects due to the presence of gold nanoparticles during PLFL with 10 ps of ZnO. Nonetheless, additional studies and effects when supported Au NP are exposed to longer laser pulses and lower laser fluences are shown in chapters 4.2.2 and 4.2.3.

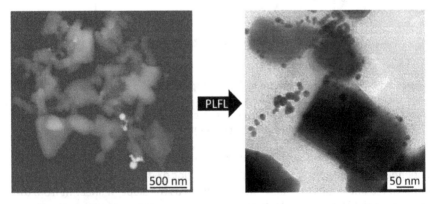

Figure 28: SEM picture of 5wt% gold nanoparticles on zinc oxide (left) and TEM picture of 5 wt% gold nanoparticles on zinc oxide after 10 passages

4.1.4 Ligand-free gold atom clusters adsorbed on graphene nano sheets generated by oxidative laser fragmentation in water

Synopsis

This chapter presents how the size limit accessible by PLFL of around 4 nm can be overcome by adding an oxidative reagent to the nanoparticle suspension during the fragmentation process. It is known that surface oxidation contributes to colloidal stability by adding surface charges causing electrostatic repulsion [Sylvestre2004], [Sylvestre2004a]. Laser fluence studies on the achievable Au NP size for ns pulses show a size limit of around 4nm [Amendola2007], [Rehbock2014]. In Fig. 29 the proposed influence of the oxidative reagent H_2O_2 is schematic illustrated. In contrast to partial reduction resulting in defect-rich zinc oxide particles the presence of an oxidative reagent contributes to preserve particle growth. The influence of hydrogen peroxide on PLFL is studied in this chapter.

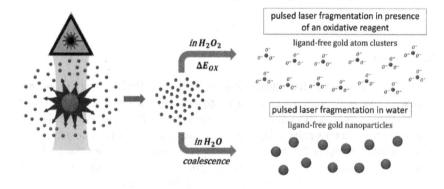

Figure 29: Schematic illustration and mechanistic conclusion for synthesis of ultra-small gold nanoparticles in presence of H_2O_2 [Lau2014d]

M. Lau, I. Haxhiaj, P. Wagener, R. Intartaglia, F. Brandi, J. Nakamura, S. Barcikowski, *Ligand-free gold atom clusters adsorbed on graphene nano sheets generated by oxidative laser fragmentation in water*, Chem. Phys. Lett. 610 (**2014**) 256-260

Abstract

Over three decades after the first synthesis of stabilized Au55-clusters many scientific questions about gold cluster properties are still unsolved and ligand-free colloidal clusters are difficult to fabricate. Here we present a novel route to produce ultra-small gold particles by using a green technique, the laser ablation and fragmentation in water, without using reductive or stabilizing agents at any step of the synthesis. For fabrication only a pulsed laser, a gold-target, pure water, sodium hydroxide and hydrogen peroxide are deployed. The particles are exemplarily hybridized to graphene supports showing that these carbon-free colloidal clusters might serve as versatile building blocks.

Introduction

Gold clusters have attracted much attention because of their optical properties. Instead of plasmonic gold nanoparticles with metallic character, ultra-small gold particles show quantization of electronic energy levels [Zheng2011, Qian2011]. The size threshold for significantly quantum-sized gold clusters can be estimated by the jellium model to be ~2 nm assuming complete defect-free filling of the valence shells [Zheng2004]. At much higher number of gold atoms, energy level spacing eventually becomes comparable to available thermal energy so that this transition size regime connects atomic and metallic behavior, as can be seen by proto plasmonic fluorescence properties. These clusters are of particular interest in a variety of application fields, e.g. as optical limiters [Philip2012], for heterogeneous catalysis [Hughes2005, Huang2009, Turner2008] or as fluorescent markers [Goldys2012, Wen2011, Shang2011], where gold clusters have already been successful applied, but till now only synthesized with reduction chemistry and stabilizing ligands. Gold particles smaller than 2-3 nm show no surface plasmon resonance in the extinction spectra and have non-linear optical properties [Lica2004, Palpant1998].

Although fluorescence of gold clusters has been shown by several groups, it is still unclear if fluorescence origins from the gold itself or the orbital coupling when organic molecules are adsorbed on particle surface [Lin2009a]. Recently, Lin et al. showed that fluorescence properties of gold clusters are strongly affected by the ligands attached to gold atom clusters surface [Lin2009b]. Beside this clusters are known to be efficient for homogeneous and especially heterogeneous catalysis [Nijhuis2006] where the catalytic activity depends on the size of the particles [Haruta1997]. Availability of colloidal gold atom clusters free of organic molecules would allow the study of optical or catalytic properties in this atom-metal transition particle size regime without cross-effects by surface absorbed ligands.

In 1993 Fojtik and Henglein reported for the first time the laser-based synthesis of inorganic nanoparticle colloids [Fojtik1993]. Since the 2000s laser ablation and fragmentation in liquids became a research field growing by factor 15 in a decade [Barcikowski2009] allowing to produce colloids of high purity via a facile synthesis route that fulfills the twelve principles of green chemistry [Amendola2013, Murphy2008a]. Gold nanoparticles, for example, are obtained simply by focusing a laser beam on a gold target placed in water without any reactants [Neddersen1993]. These gold nanoparticles show a high reactivity due to their pure (ligand-free) surface and can be used, e.g. for bio-conjugation with a variety of molecules [Barchanski2012, Barchanski2011, Petersen2011] or sorption on supporting particles [Wagener2012a]. Also bimetallic nanoparticles colloids can be fabricated by laser ablation and fragmentation in liquid, e.g., Silver/Gold [Intartaglia2013]. As a drawback, laser-generated particles have a wide size distribution compared to chemically-synthesized ones [Besner2006, Besner2010]. To reduce this scientific drawback of polydispersity, size quenching effects by adding salts or ligands [Rehbock2013] can be used, resulting in monodisperse gold colloids. An alternative size reduction method is post irradiation with ultra-short laser pulses avoiding contaminations by additives [Menendez-Manjon2011, Amendola2007, Werner2011].

It is known that pulsed laser irradiation of gold colloids with laser fluence of 0.05-0.5 J/cm² may cause a reduction of particle diameter down to a size limit of 4 nm [Riabinina2011]. Yet, only the addition of organic surfactants gave access to particle sizes smaller than 4 nm via the laser fragmentation route [Giammanco2010, Mafuné2002], but unfortunately the employed stabilizers cover the final gold cluster surface and are difficult to remove quantitatively.

Materials and Methods

Laser ablation and fragmentation

In our experiments, we firstly generated educt particles by laser ablation of a gold target in an ablation chamber filled with deionized water (Millipore pH ~ 5-6) using a picosecond pulsed Nd:YAG laser (Ekspla, Atlantic Series, 10 ps, 100 kHz, 150 μJ, 1064 nm) for 5 minutes yielding ~ 150 mg/L gold nanoparticles. This suspension was diluted to a concentration of 10 mg/L. The subsequent laser fragmentation procedure was performed using the second harmonic of a Nd:YAG nanosecond laser (Innolas, Spitlight, 10 ns, 100 Hz, 75 mJ, 532 nm) for 4 hours with a fluence of about 0.7-0.8 J/cm² without (pH 8) and with 10 wt-% hydrogen peroxide (Fluka) in water (pH 5), and 0.6 mM NaOH (Applichem). The experiments were performed using the 4.5 mm raw beam in a glass vessel with 4 mL gold nanoparticle suspension.

Graphene nano sheets and adsorption of gold atom clusters to GNS

The graphene nano sheets were prepared by the reduction from oxidized graphite. This oxidized form of graphite was dispersed in water and reduced by hydrazine hydrate, resulting in graphene nano sheets as described from Siburian et al. [Siburian2013]. The gold atom clusters used for this experiment were fabricated in 10 wt-% hydrogen peroxide. Subsequently these gold clusters were adsorbed to the GNS under continuous treatment with a sonotrode and by heating of the cluster suspension to 80°C. GNS fabrication and gold cluster adsorption were performed in the Faculty of Pure and Applied Sciences at the University of Tsukuba.

Particle Analysis

Gold particle suspensions were characterized by UV-vis spectroscopy using a Cary 50 (Varian Inc.) spectrometer. TEM pictures were taken with a Zeiss EM190 microscope and a Jeol JEM-2100 high resolution transmission electron microscope (NIMS, Japan). Particle sizes were detected in an analytical disc centrifuge DC 24000 (CPS instruments) at 24000 rpm and by dynamic light scattering (DLS) using a Zetasizer Nano ZS (Malvern).

Results and Discussion

Here, we present a combination of laser ablation and oxidative fragmentation to generate ligand-free colloidal gold clusters. Laser fragmentation of gold nanoparticles in pure water results in a particle size reduction similar to experimental findings of Amendola et al. [Amendola2007]. Size quenching and stabilizing effects for small amounts of salts are currently reported [Merk2014, Pfeiffer2014]. Here we combined the stabilizing effect of a small amount of sodium hydroxide (NaOH) with the oxidative potential from hydrogen peroxide (H_2O_2). Both additives are chosen since they offer the opportunity of a carbon-free stabilization of gold particles < 4 nm. Addition of NaOH is not mandatory, but helpful to increase the pH of the solution for sufficient colloidal stability as the low pH resulting from the addition of H_2O_2 could cause destabilization of laser-generated colloidal gold particles. Nevertheless, all involved species are based on oxygen and hydrogen and no irreversible attachment of anions like chloride have to be considered. For applications where sodium might be harmful (e.g. specific catalytic reactions) the cation could be replaced by ammonium or other alkaline additives.

The corresponding UV-vis spectra of the particle colloids after ablation and for fragmentation with 0.6 mM sodium hydroxide show a shift in plasmon resonance from 530 nm to 510 nm (Figure 30) which is equivalent to a particle size of around 40 nm and 6 nm respectively [He2005, Haiss2007]. In contrast to this, the presence of hydrogen peroxide during the fragmentation process changed the particle size of the colloid without observing any precipitation. The significant decrease of plasmon resonance of the fragmentized particles might be explained by the fabrication of non-plasmonic particles [Garcia2005] in both cases, with and without hydrogen peroxide. However, if hydrogen peroxide is present the particles size further decrease. This is observed by a plasmon resonance peak shift from 530 to 507 nm.

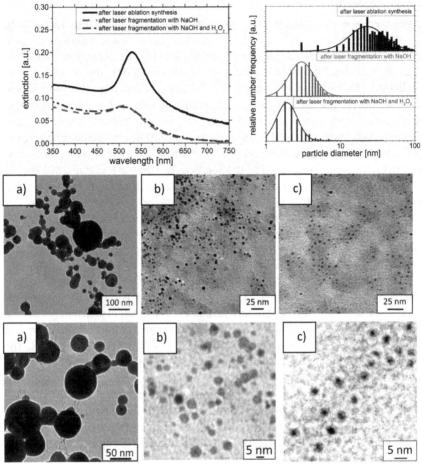

Figure 30: UV/Vis-spectrum and size distributions of gold nanoparticles produced by pulsed laser ablation and fragmentation. Analysis was performed after laser ablation (black curve, top), fragmentation of particles in water with sodium hydroxide (red curve, middle) and after oxidative laser fragmentation in presence of sodium hydroxide and hydrogen peroxide (blue curve, bottom). Particle size distribution is determined from TEM images, which are exemplary shown on the bottom (a: after laser ablation, b after pulsed laser fragmentation in water with low NaOH concentration (0.6 mM), c after fragmentation in presence of hydrogen peroxide (10 wt-%) and 0.6 mM NaOH.

Figure 30 summarizes the influence of hydrogen peroxide on the particle size. Analysis was performed after laser ablation (black curve, top), fragmentation of particles in water with sodium hydroxide (red curve, middle) and after oxidative laser fragmentation in presence of sodium hydroxide and hydrogen peroxide (blue curve, bottom). Particle size distribution is determined from TEM images, which are exemplary shown on the bottom (a: after laser ablation, b after pulsed laser fragmentation in water with low NaOH concentration (0.6 mM), c after fragmentation in presence of hydrogen peroxide (10 wt-%) and 0.6 mM NaOH.) The plasmon resonance shift and decrease of extinction in the UV/Vis-spectra correlate to the decrease of particle size after fragmentation which is also observed from TEM images.

This size reduction was, confirmed by transmission electron microscopy (TEM) images and analysis in an analytical disc centrifuge (see also Figure 8-13 – 8-15 and Figure 8-18 – 8-20). Gold nanoparticles generated by laser ablation without surfactants usually obtains particle diameter between 10 and 100 nm (Figure 30). Laser fragmentation reduces the size of these particles down to a mean size of 3.5 nm, for minute amounts of sodium hydroxide. TEM-analysis of particles fragmentized in presence of both sodium hydroxide and hydrogen peroxide show product particles smaller than 3 nm whereas most particles are around 2.1 nm. These observations in the difference when hydrogen peroxide as oxidative reagent is present during the fragmentation process are statistically confirmed by particle size analysis using an analytical disc centrifugation, reported in Figure 31.

This quantitative analysis confirms reduction of the particle size after the fragmentation and the positive effect of hydrogen peroxide for further particle size decrease. Therefore: i) the educt gold nanoparticles are in a regime of tens of nanometers; ii) laser fragmentation in water with 0.6 mM sodium hydroxide results in a hydrodynamic particle diameter of around 10 nm; iii) the presence of hydrogen peroxide yields around 4-6 nm particles.

Deviations of TEM histograms and the results from the analytical disc centrifuge might be explained by the measurement method. Sizes observed in the TEM correspond to a real particle diameter whereas particle sizes resulting by analytical disc centrifugation are enhanced by contributions of soft agglomerates and solvation shell.

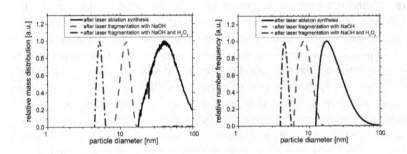

Figure 31: Hydrodynamic particle diameter (mass distribution and number frequency) of gold nanoparticles after laser ablation (black/solid curve), fragmentation of particles in water with sodium hydroxide (red/dashed curve) and after oxidative laser fragmentation in presence of sodium hydroxide and hydrogen peroxide (blue/dased and dotted curve) respectively, measured in an analytical disc centrifuge (ADC).

Figure 32 (right) shows a zoomed gold cluster from a HR-TEM image with a measured lattice spacing of 0.2 nm (illustrated as black lines and arrows) which correspond to gold {200} planes in the Au lattice direction. Therefore, laser fragmentation in presence of both sodium hydroxide and hydrogen peroxide is capable to generate smaller nanoparticles than fragmentation with sodium hydroxide alone. It is known that laser fragmentation in liquid is an ultrafast process on a picosecond timescale [Ibrahimkutty2011] therefore the observed influence of oxidative species should be related to a subsequent stabilization of laser fragments instead of an impact on the fragmentation mechanism itself. We believe that this oxidative species is capable to stabilize the oxidized nanoparticle surface resulting in charged surface atoms that contribute to electrostatic stability. Muto et al. have shown that a significant amount of surface atoms of laser-generated nanoparticles is oxidized to Au+ or Au3+ species [Muto2007].

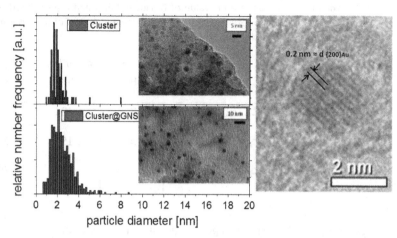

Figure 32: Histograms of particle sizes from laser-fabricated gold clusters (in pure hydrogen peroxide) before (top) and after (bottom) adsorption on graphene nano sheets on the left with HR-TEM images as inserts and a HR-TEM micrograph of a gold cluster at the right side.

These species are usually covered by anions or other negatively-charged species that give a pure electrostatic stability to the laser-generated nanoparticles with high surface charge density (see characterization of the isoelectric point of gold particles within the Supporting Information, Figure 8-21). Consequently, nanoparticle surface oxidation accounts for the colloidal stability of laser-generated nanoparticles. In case of laser fragmentation, size reduction to smaller nanoparticles or even clusters strongly enhances the total nanoparticle surface of the system. These particulate fragments underlie coalescence and ripening processes and are growing to bigger particles that limit minimal particle size obtained in equilibrium by laser fragmentation in pure water. Inside diluted electrolytes there is size-quenching [Rehbock2013] effect shown by effective particle size reduction for fragmentation in diluted sodium hydroxide.

In presence of an oxidizing species, these clusters might be stabilized additionally by oxidation contributing to surface charge density and electrostatic interparticle repulsion hindering coalescence. Therefore the presence of hydrogen peroxide and minute amounts of sodium hydroxide are capable to fabricate ultra-small gold nanoparticles.

Table 3: Oxidation potential at a pH value of 0 [Lide2009], 5 and 8 (calculated by the Nernst equation) for oxygen, hydrogen peroxide and Au3+. Potential is given for the pH values used for the experiments.

	$E°$ / V (pH 0)	$E´$ / V (pH 5)	$E´$ / V (pH 8)
$H_2O_2 + 2H^+ + 2e \rightleftharpoons 2H_2O$	1.776	1.481	--
$Au(OH)_3 + 3H^+ + 3e \rightleftharpoons Au + 3H_2O$	1.45	1.155	0.978
$O_2 + 4H^+ + 4e \rightleftharpoons 2H_2O$	1.229	0.934	0.757

This process of oxidative laser fragmentation enables quantitative size reduction of gold nanoparticles and results in the fabrication of colloidal gold atom clusters in the quantum-size regime without using any organic ligands.

Looking at the redox potential is helpful to identify oxidizing species able to stabilize gold particle surface sufficiently. Table 1 gives the redox potential of dissolved molecular oxygen, hydrogen peroxide and bulk gold at given pH 5 and 8. As can be seen, the redox potential of hydrogen peroxide is slightly higher compared to Au3+ which might result in enhanced surface oxidation and subsequent stabilization of the gold clusters. Even though the oxidation potential for nanoparticles was reported to decrease close to the oxidative potential of O2 if the nanoparticle size decreases [Ivanova2010], it becomes difficult to oxidize clusters surface by water-solved O2 in a sufficient way. As can be seen the oxidative potential of hydrogen peroxide is above the potential of gold, whereas the potential of oxygen is lower. Thus oxygen present in the solution will not be able to oxidize gold particles surface whereby hydrogen peroxide can. Further we calculated the molecules stoichiometrically available for each surface atom in case of an oxygen saturated water solution at 20°C and for 10 wt-% hydrogen peroxide solution for a gold concentration of 20 mg/L. This diagram, shown in the supporting information (Figure S4.1.4 1) depicts that only few molecules of oxygen are available for the gold surface even if the water is saturated with O2. Therefore, sufficient surface oxidation by O2 is kinetically improbable even when the oxidative potential for gold nanoparticles and clusters is close to the value of molecular oxygen. In contrast to this, the oxidative potential of hydrogen peroxide is significantly above the value for gold, and H2O2 is available in excess by 4 orders of magnitude, therefore delivering both the necessary potential difference and molecular quantity, not provided by oxygen.

Furthermore, we adsorbed the ligand-free gold clusters after the synthesis on graphene nano sheets (GNS) (Figure 32, see also Figure 8-16 and 8-17) showing that these clusters might be used as building blocks for heterogeneous catalysts [Mafuné2014, Siburian2012]. Note that these clusters were fabricated only in presence of hydrogen peroxide. No deviation of particle size distribution before and after adsorption on GNS is observed and the resulting hybrid material may open up catalytic applications which highly demands very small particles or even clusters. As all observed clusters are located on the GNS support (see Figure 32 and Figure 8-17 Supporting Information) and no separated particles or clusters are observed, we expect a sufficiently strong interaction between clusters and support. In literature, the adsorption of clusters and nanoparticles to graphene or graphene-like surfaces is usually assigned to van der Waals forces [Rance2010]. Additionally, Coulomb interactions with Au particles may be considered at the defects sites of graphene [He2013]. To estimate the strength of gold cluster adsorption it is helpful to consider the Hamaker constant, which describes a pair potential between nanoparticles and carbon support. In case of metal nanoparticles and carbon nanotubes (that are similar to the used graphene nanosheets), the value of this constant is quite large (6×10^{-19} J, taken from Akita et al. [Akita2000]). This indicates a strong interaction between clusters and graphene surface. Consequently, the adhesion of gold atom clusters to graphene nanosheets should be irreversible and strong enough even to perform rinsing steps in order to purify the heterogeneous catalyst.

Recently, Mafuné et al. showed the laser-fabrication of ultra-small nickel particles by in-situ size quenching in presence of silica nanoparticles [Mafuné2014]. However, this one-pot method is restricted to fine dispersed supporting materials without any absorption or scattering properties at used laser wavelength. By the presented two-step method (laser fragmentation and subsequent supporting) almost any supporting material can be used.

Conclusions

In summary, ligand-free gold clusters with particle sizes smaller than 3 nm may be synthesized in water during pulsed laser fragmentation of the laser-synthesized gold nanoparticles in presence of sodium hydroxide and hydrogen peroxide as oxidative reactant. The gold clusters can be quantitatively adsorbed on graphene with potential use as heterogeneous catalyst. The availability of ultra-small gold particles in the atom-metal transition size regime that are free of any organic molecules on the cluster surface could be an interesting reference material for fundamental studies on optical or catalyst properties of quantum-sized metal colloids.

ACKNOWLEDGEMENTS

We thank Jurij Jakobi for the transmission electron microscope measurements and Akasu Yuta for the generation of graphene nano sheets.

4.1.5 Laser-induced chemical conversion of particle suspensions

4.1.5.1 Fragmentation of copper compounds – laser-induced reduction

A cooperative work combining copper-based materials used in Schaumberg et al. [Schaumberg2014] with the method of Lau et al. [Lau2014a] also aims to understand whether processing in the liquid jet affects particle composition. This chapter addresses PLFL mechanisms and the influence of the surrounding liquid and educt particle composition on obtained particle properties. For this purpose, the copper compounds copper nitride (Cu_3N), copper-(II)-oxide (CuO), copper-(I)-oxide (Cu_2O) and copper iodide (CuI) were used as educt microparticles and exposed to previously-determined optimized laser irradiation parameters with 532 nm, 10 picoseconds, 100 kHz and 75 µJ pulse energy. As liquid, ethyl acetate (EA) was chosen. Cu_3N was chosen to be fragmentized in water and also under nanosecond UV laser irradiation.

Combination of both scientifically reported experiments [Schaumberg2014], [Lau2014a] may enable further insight, and additional copper compounds are investigated. Copper iodide plays an important role here, as described in detail in the following. Figure 33 presents photographs of the dry and in EA suspended educt powder materials (left). Due to large educt particle size in the range of several micrometers, they sediment within a few minutes, leaving the pure solvent as a supernatant. After fragmentation, the nanoparticles generated from microparticles form a stabile colloid. Nanoparticle colloids were decanted after sedimentation of the microparticles, whereby the photograph on the right in Figure 33 depicts these stable colloids. In the following, the observed change of chemical composition will lead to hypothesis of fragmentation mechanism and the educt particle property influence.

Figure 34 illustrates particle size distributions of generated colloids shown in Figure 33 on the right. The histograms were taken from TEM images. Representative TEM images are shown in Figure 35 (images were made by C. Schaumberg, December 2014 at the Helmholtz-Zentrum Berlin). Additional TEM images of the different nanoparticle materials are added in supplementary data 4.5.

It is obvious that the largest nanoparticles and widest particle size distribution are obtained from copper iodide microparticles. Electron energy loss spectroscopy (EELS) of the different colloids indicates the chemical composition of generated nanoparticles. Figure 36 summarizes EELS results for the different colloids showing relevant energy regimes for copper, oxygen and/or nitrogen. If a shift in energy is observed, this results in an energy loss edge and thus signal intensity increases [Schaumberg2014]. Given that all elements show distinctive energy levels, EELS enables identification of elements. The most distinctive reduction to copper nanoparticles was obtained by PLFL of Cu_3N in EA. For CuO in EA, only a small amount of oxygen is present. This might be explained by oxygen being present in the educt microparticles, resulting in residual oxygen species after vaporization and nucleation. This would lead to the conclusion that Cu_2O in EA will show less oxygen present in nanoparticles compared to CuO, although this is not the case.

The fragmentation of CuI particles also show this correlation between particle size distribution and change of chemical composition, which is even more significant for copper iodide. CuI probably follows the shock wave-induced fragmentation mechanism as reported for the majority of ZnO in 4.1.1. This can be observed in gradual size reduction and an overall wider particle size distribution and particle shape. Additionally, CuI is not reduced to Cu, which can be another indication for the fragmentation mechanism. The preservation of the educt's elemental composition and crystallinity is confirmed by the diffraction pattern (Figure 37) and EELS (Figure 36).

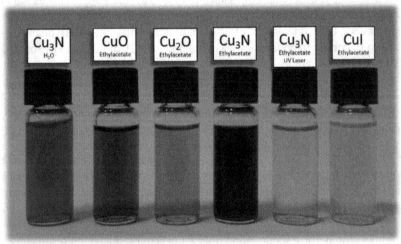

Figure 33: Photograph of the different copper microparticle compounds before (top) after (bottom) laser fragmentation (100 passages at picosecond laser (532 nm wavelength, ~30 mJ/cm²) 50 passages at UV laser (355 nm wavelength, ~3 J/cm²)) forming stable colloidal nanoparticle suspensions

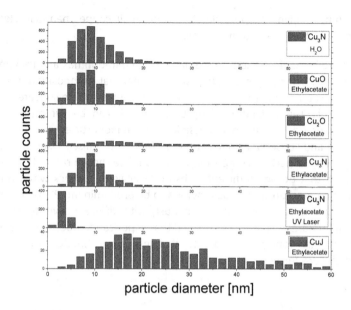

Figure 34: Histograms taken from TEM pictures of the different colloids after PLFL
shown in Figure 33 and 35

Comparing PLFL of Cu_3N in EA and in H_2O, oxidation subsequent to NP generation
is observed, confirmed by EELS (presence of oxygen shoulder) and TEM (shell
covering particles). The abundance of nitrogen confirms that particles undergo a
reduction to elemental copper. It might be assumed that particles only have an
oxidized shell and might be $Cu@CuO_x$ but cannot be confirmed by data. TEM
images only indicates this by the low contrast shell that surround particles. In
supplementary data 4.5, several additional TEM images are shown, where a
crystalline structure can be observed for PLFL of Cu_3N in EA.

Interesting results were found for ns PLFL of Cu_3N in EA with 355 nm, where
particles are found to be very small. The color indicates copper acetate upon first
glance, although TEM reveals small particles. Elzey at al. reported colloial copper
nanoparticles formed in aqueous HCl [Elzey2011]. If copper is dissolved in HCl, it
appears to be bluish in color, similar to our Cu_3N colloid, shown in Figure 33
(bluish liquid). This might indicate that Cu_3N was dissolved in EA by laser
irradiation. Given that TEM images revealed small nanoparticles with diameters

below 3 nm, it might be possible that very small copper nanoparticles are generated, which appear bluish in color.

The surrounding liquid present during laser fragmentation has a crucial impact on obtained particles, besides the starting material composition used. In presence of water nanoparticles generated from copper nitride indicate to have an oxide shell. This is confirmed by EELS. Thus it can be summarized that reductive laser fragmentation is possible when nanoparticles are formed by vaporization and subsequent nucleation, as known for pulsed laser ablation in liquid. To preserve particle composition, EA should be chosen rather than water. For particles that do not sufficiently absorb laser light, a shock wave-induced fragmentation mechanism can be assumed to be predominant. Please note that both mechanisms could occur simultaneously with different severity, as found for zinc oxide. Here, it cannot be excluded that vaporization does not take place, although if this vaporization occurs it might be estimated that copper nanoparticles are formed.

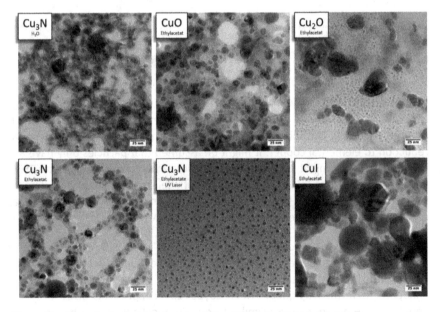

Figure 35: Representative TEM images of the different colloids after laser fragmentation

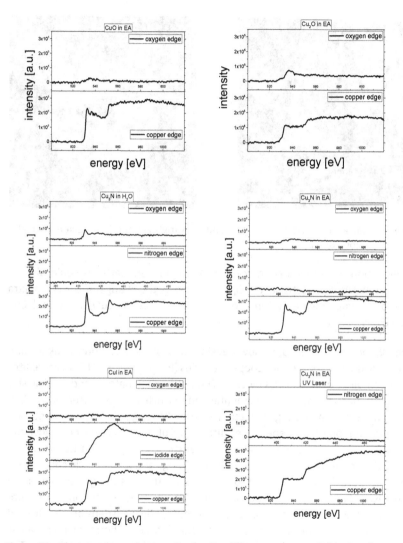

Figure 36: Electron energy loss spectra for the different copper colloids showing presence of copper and the corresponding relevant elements, such as oxygen, nitrogen or iodide

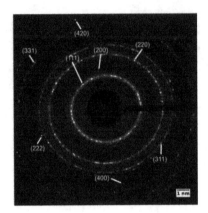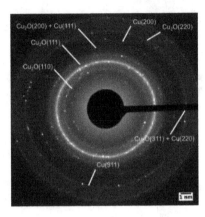

Figure 37: Diffraction pattern of CuI (left) and Cu₃N (right) after laser fragmentation at picosecond laser

The evolution of the extinction spectra of copper nanoparticles could also be observed by UV-vis spectroscopy. Figure 38 shows the change of UV-vis spectra evolution during fragmentation process in EA for Cu_2O, CuO, Cu₃N and CuI, respectively.

In Fig. 38, only CuO and Cu₃N show development of the plasmonic peak around 570 nm, distinctive for metallic copper particles [Pestryakov2004]. This correlates with the findings from analysis by EELS, TEM, diffraction pattern and particle size distribution histograms where elemental nanoparticles can be observed. All UV-vis spectra show increasing extinction in the UV range during fragmentation confirming nanoparticle generation, although CuI and Cu_2O do not show the development of a plasmon peak characteristic for copper particles. UV-vis spectra underline the previously described findings and mechanistic conclusions regarding particle absorption properties and the resulting fragmentation mechanisms. For copper (II) oxide, an additional experiment was conducted with argon as ambient protection gas. For this purpose, educt suspension was flushed with argon for 30 minutes prior to laser fragmentation. The storage container and liquid jet were flushed with argon continuously during the experiment. The corresponding UV-vis spectra show a more distinctive development of the surface plasmon peak for copper, as shown in Figure 39 and supplementary data Figure 8-31. This means that dissolved oxygen species have a crucial impact on oxidation during PLFL.

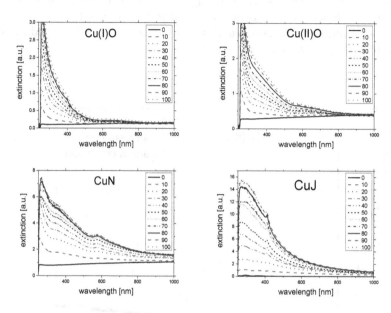

Figure 38: UV-vis spectra of the different copper compounds fragmentized under optimized conditions in ethyl acetate up to 100 passages

To summarize these important findings for the mechanistic understanding, it can be stated that laser fragmentation mechanisms depend on educt particle properties. From the experimental findings shown here for different copper compounds, as well as the observations for the fragmentation of ZnO and B₄C, it might be assumed that absorption efficiency is crucial for mechanisms occurring; this might also define whether a change in chemical composition is possible. In the case of CuI and ZnO particles, a shock wave-induced fragmentation mechanism reducing particle size gradually appears to be predominant. If the absorption of particles is high, laser light couples more distinctively and thus vaporizes them more sufficiently to cause a chance of chemical composition. Smaller particles might be copper nanoparticles and are spherical. If the vaporization mechanism is predominant (as for copper nitride), the obtained nanoparticles are produced directly (likewise for PLAL) from microparticles. Here, the surrounding liquid plays an important role for the final composition of product particles.

This offers an interesting insight into the sensitivities of PLFL against educt particle properties and the surrounding liquid. The controlled reduction to elemental materials can be attenuated, as shown for PLFL in water. In contrary, the issue of whether controlled oxidation can be achieved by PLFL is addressed in the following chapter 4.1.5.2.

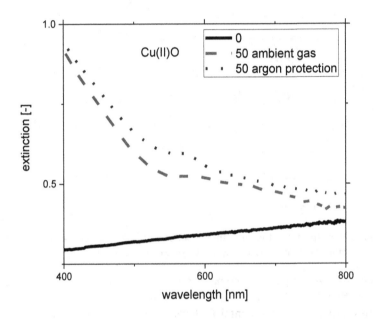

Figure 39: UV-vis spectrum of CuO fragmentized under ambient gas and under argon gas protection

4.1.5.2 Fragmentation of aluminum particles – laser-induced oxidation

Besides the reduction of copper compounds, the fragmentation of elemental aluminum is investigated. Aluminum particles were fragmentized for 50 passages under optimized fragmentation conditions found for zinc oxide before (532 nm laser wavelength, 100 kHz repetition rate, ~30 mJ/cm²). Figure 40 a) is a photograph of the colloids before and after laser fragmentation, depicting the impact of laser irradiation on oxidation of aluminum particles. After storage of 66 hours fragmentized particles appear to be completely oxidized. This laser-induced hydrolysis reaction was confirmed by gas chromatographic measurements analyzing the ambient gas above the colloids. This intense reaction produces a significant amount of H_2, which could be observed by bubble formation in the liquid. By inflaming the gas above the stored colloid, an oxyhydrogen reaction could be induced, confirming the presence of hydrogen. Interestingly, this reaction starts approximately 20 hours after laser irradiation. ADC data is shown in Figure 40 b) show particle size distributions of aluminum particles before and after laser irradiation. A strong aggregation of generated nanoparticles is observed due to the absence of any stabilizing agents (see Figure 40 b) red curve). This confirms nanoparticle fabrication from microparticles, as well as that particles are not stable.

In supplementary data Figure 8-32, time-resolved photographs of the particles are shown, confirming enhanced oxidation after fragmentation of aluminum particles. This can be related to the higher specific surface area generated by fragmentation. Additionally, water is required to cause this oxidation, as confirmed by reference experiments of aluminum particles fragmentized in EA, ethanol and isopropanol (see Figure 8-33).

An enhanced absorption of the laser fragmentized particles in water after 50 passages can be observed by UV-vis spectra. During oxidation, extinction in the visible to UV regime increases. Corresponding spectra after 0, 43 and 66 hours of storage are shown in Figure 41.

All organic solvents used (EA, ethanol and isopropanol) prevented distinctive oxidation. Aluminum nanoparticles were observed to be stable in ethanol and isopropanol, respectively. Please note that no report on hydrogen fabrication subsequent to laser fragmentation or enhanced oxidation of aluminum particles by laser fragmentation in liquid environment exist to date.

Gas chromatograph analysis of the ambient gas above colloids were performed for un-fragmentized particles (educt), fragmentized particles (50 passages) after approximately four weeks of storage, respectively. To confirm a retention time of 8.22 minutes for hydrogen, a hydrogen testing gas as a reference is shown in Figure 42. This diagram depicts the fragmentized sample, the reference, hydrogen as testing gas, as well as the laboratory gas (from top to bottom).

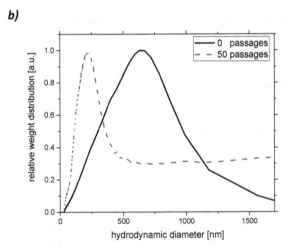

Figure 40: a) Aluminum particles in water after 66 hours. Particles in the left vessel
were not fragmentized (reference), particles in the right vessel were
fragmentized 50 passages under optimized conditions; b) Particle size
distribution as measured by ADC of aluminum particles before and after 50
passages in water

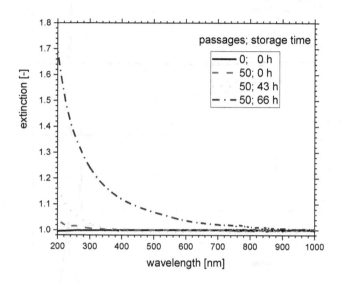

Figure 41: Extinction spectrum of aluminum particles irradiated 50 passages after different storage times after laser activation

Oxidation of laser fragmentized particles started around 20 hours after laser activation (PLFL size reduction) and the suspension appeared to be sufficiently oxidized after around 66 hours. The area of the hydrogen peak detected around four weeks after suspending particles in water and fragmentizing them (reference particles were suspended solely) is higher than the reference by a factor of 3. Hydrolysis of particles' surface also takes place for un-fragmentized particles, albeit on a significantly slower and less distinctive basis. Further investigations regarding time resolved hydrogen production will provide a more detailed insight into the PLFL-based activation (size reduction) for hydrogen production.

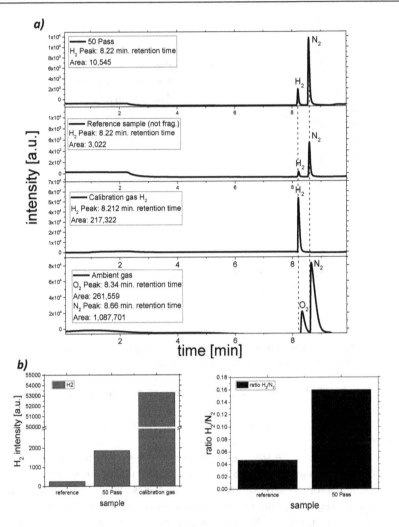

Figure 42: a) Gas chromatograph analysis of the headspace developed above the aluminum particles suspended in water after approx. 100 hours; b) bar chart of the H_2 intensity signal (red bars, left) and the ratio of H_2 to N_2 (black bars, right)

Here, it could be shown that enhanced hydrolysis takes place after laser fragmentation of particles. This oxidation is attenuated if organic solvents such as isopropanol, ethanol or EA are used. Fragmentation of aluminum particles in isopropanol or ethanol result in nanoparticle colloids that do not precipitate.

It could be shown that laser activated aluminum particles oxidize significantly and release hydrogen from water environment, whereby particles are "activated" by laser irradiation. It can be assumed that for this the specific surface area plays an important role. Note that the formation of aluminum and defect-rich aluminum oxide nanoparticles from PLAL of an aluminum target in water is reported [Kumar2010], although not the hydrogen release subsequent to PLFL of aluminum particles.

4.2 Laser melting

4.2.1 Pulsed laser melting in liquids (PLML) for the fabrication of sub-micrometer spheres (SMS)

Pulsed laser melting in liquid (PLML) is a technique to generate sub-micrometer spheres (SMS) from micro-, sub-micro- or nanoparticles. Similar to laser fragmentation, particles are exposed to laser light and consequently change their size or shape; however contrary to PLFL, the size does not decrease during PLML.

Koshizaki and co-workers first reported the formation of boron carbide SMS by PLML of boron particles in EA [Ishikawa2007]. They proposed a mechanism of the chemical conversion for B_4C SMS formation and discussed the impact of laser fluence on obtained particle size and the resulting boron carbide yield from boron particles [Ishikawa2010]. For PLML they focused a 355 nm ns laser beam into a vessel. Thus, they achieved different laser fluences in the reaction vessel during beam propagation. They investigated different effects due to different mechanisms occurring in the different corresponding fluence regimes, whereby their mechanistic conclusion was the formation of B_4C particles (chemical conversion) in the high fluence regime and PLML to SMS in the lower fluence regimes.

Cai and co-workers recently reported about the precise control of sphere formation using non-spherical gold nanoparticles for SERS enhancement [Liu2015]. This precise control was possible due to use of an unfocused laser beam, thus avoiding fluence gradient along beam propagation through the vessel. Hence, for both dielectric and metal particles, precise fluence control and minimized beam divergence or thin liquid layer is important during PLML.

This chapter reports on the sequential laser irradiation in a passage reactor to investigate PLML with precise laser fluences. Experimental procedure was adapted from the experiments for PLFL (chapter 4.1.1), thus 0.1 wt% of particles were suspended and irradiated in surfactant-free 50 ml deionized water and stabilized in 0.1 M SDS solution after the last passage in a ratio of 1:1 (no SDS present during PLML processing).

Figure 43 shows SEM images of zinc oxide particles before (left) and after (right) PLML. Images indicate that elongated ZnO crystals melt and form SMS. Note that SMS are crystalline as confirmed by XRD spectrum of particles before and after melting, shown in Figure 44 together with the corresponding UV-vis spectra and ADC measurements.

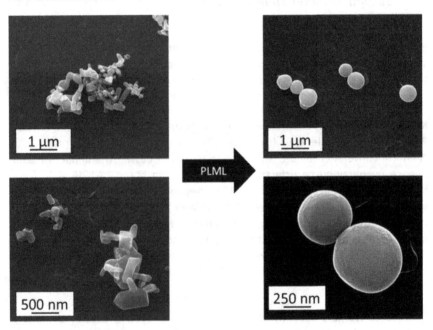

Figure 43: Representative SEM images of zinc oxide particles before (left) and after (right) ns PLML at 355 nm laser wavelength with ~100 mJ/cm²

UV-vis spectra show a bathochromic shift of extinction peak in the visible range that is caused by scattering of generated SMS [Stamatakis1990]. This obtained peak shift from scattering of SMS can be used to describe the PLML efficiency for ZnO particles.

Plotting the obtained peak shift in visible wavelength region versus the applied UV-laser fluence shows that also by applying UV irradiation fragmentation can occur (Figure 45). In the high fluence regimes a peak shift around 0 means that no melting but rather fragmentation occurs. Increasing the fluence by adapting the liquid jet directly into focal plane did not result in an optical breakdown of the pure liquid (no particles suspended).

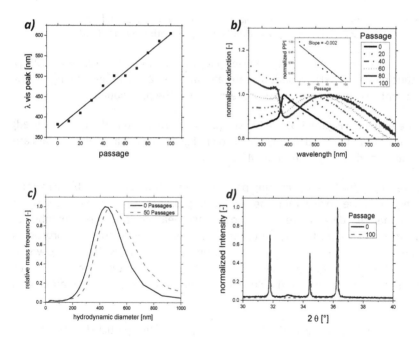

Figure 44: Shift of visible peak from UV-vis spectra (a)), UV-vis spectra after increments of 20 passages (b)), hydrodynamic diameter from ADC results before and after 50 passages (c)), and XRD pattern before (0 passages) and after 100 passages irradiated with 355 nm laser light (d))

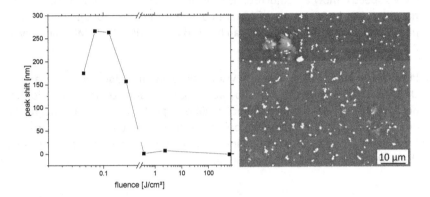

Figure 45: Influence of fluence on UV-laser PLML efficiency (peak shift) after 50
passages and SEM image after 50 passages with around 100 mJ/cm²

Thus it can be assumed that the laser parameters of 40 ns pulse length, 23 W output power, 85 kHz repetition rate and approx. 270 µJ pulse energy is not sufficient to cause a visible optical breakdown of the pure liquid.

In the Appendix Figure 8-37, all UV-vis spectra (normalized and non-normalized) for the different applied fluences are shown (figures from that peak shift data are extracted and displayed in Figure 46).

Interestingly, for fluences causing particle fragmentation (> 120 mJ/cm²) strong change is observed within the first ten passages (see UV-vis spectra in supplementary data Figure 8-37). Plotting the peak position versus number of passages (Figure 46) for the different laser fluences shows that this process is approaching saturation. This is also confirmed by SEM images (Figure 45), where (almost) all observed particles are of spherical shape after 50 passages. For fluences between 200 and 380 mJ/cm², no distinctive particle melting is observed.

Instead, a slight increase of the PPI is observed, indicating particle fragmentation. This effect is most distinctive within the first ten passages. Figures 48 and 52 a) demonstrate that the passage reactor is an appropriate technique to determine process windows and control the laser fluence precisely during PLML.

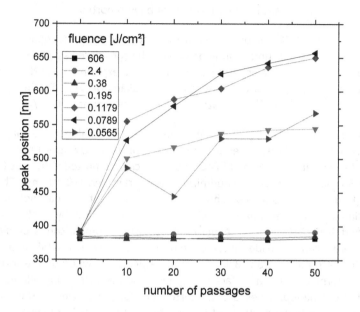

Figure 46: Vis extinction spectra peak position for different fluences plotted versus number of passages for 0.1 wt% zinc oxide particles irradiated with 355 nm, 40 ns

In case this method could be advanced to material combinations, the method of PLML could reflect enormous potential for compounding materials to generate hybrid SMS that are not otherwise accessible. Exposing supported gold nanoparticles onto zinc oxide particles [Lau2014e], [Marzun2014], [Wagener2012a] to PLML the question arises, whether gold nanoparticles can be embedded into ZnO forming hybrid SMS by PLML. At present, no reports on nanoparticles embedded into ceramic SMS by PLML are available. Parallel to these experiments Nakamura et al. reported calcium phosphate SMS with embedded iron nanoparticles (magnetite) [Nakamura2015]. They used chemical precursor solutions containing calcium, phosphate and iron ions. By exposing this precursor solution to 355 nm laser irradiation with fluences of 67-200 mJ/cm², they synthesized SMS containing Fe NP with different magnetic properties. The approach reported can be regarded as bottom-up laser-induced SMS particle synthesis, although a synthesis of such multicomponent SMS by PLML of precursor-free particle suspensions has not yet been reported.

Hence, PLML of ZnO microparticles with supported gold nanoparticles is investigated for the first time. This approach can be considered as a "reshaping" procedure [Wang2010], [Wang2011] and differs from experiments of Nakamura et al., given that we did not use dissolved ions [Nakamura2015]. The impact of the presence of Au NP was investigated for fluences of around 120 mJ/cm².

Here, different amounts of gold nanoparticles (prepared analogous to chapters 4.1.3 and 4.2.3) were suspended in water and exposed to UV laser irradiation. Figure 47 shows an SEM image of ZnO microparticles supported with 5 wt% Au NP after 50 passages exposed to optimized PLML parameters (80 mJ/cm²). These images (Figure 47, bottom) were chosen because it presents a "snapshot" of an un-irradiated educt microparticle agglomerate surrounded by molten and resolidified SMS. Thus a direct educt-product particle comparison is possible. Images were taken of the identical section by a secondary electron (SE) detector and back scattered electron (BSE) detector, respectively, whereby the former provides information about surface topography and the latter about elemental contrast. Gold nanoparticles are supported on educt particles' surface and appear to be embedded after PLML processing. Thus, PLML of micro/nano particle compounds enables hybridization and embedding of nanoparticles into the matrix of SMS. Note that 5 wt% Au NP on ZnO is equivalent to 1.5 vol% due to the difference of around 3.4 in bulk materials' density. In Figure 8-11, a diagram showing this mass-volume correlation for gold supported to zinc oxide is drawn. Laser excitation is proportional to volume fraction of Au NP, rather than mass.

From the key parameters peak shift, slope of PPI or obtained particle size distribution, a slight influence on the process efficiency by the presence of Au NP was observed. Figure 49 compares UV-vis spectra after 50 passages for optimized PLML parameters with a different amount of gold nanoparticles. This might indicate that the presence of gold nanoparticles results in a lower peak shift. Considering the laser light consumption of Au NP could explain this. Nevertheless, it is possible to generate Au NP/ZnO SMS hybrid particles by PLML.

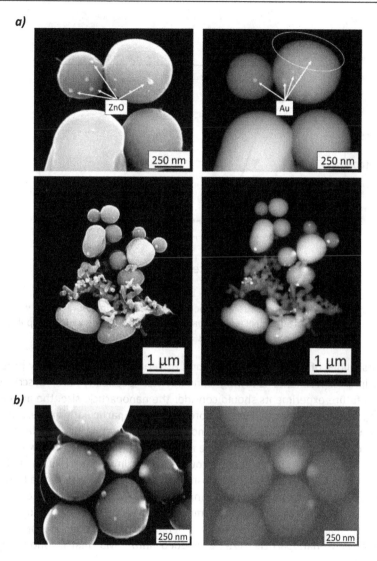

Figure 47: a) Zinc oxide microparticles with 5 wt% gold nanoparticles irradiated at 355 nm laser wavelength (80 mJ/cm², 40 ns) detected with secondary electrons (left) and back scattered electrons (right), on top a magnification of particles is shown with indication of gold and zinc oxide nanoparticles on and into the sub-micrometer sphere respectively b) Product of PLML of ZnO-supported Au NP, SEM images of zinc oxide sub-micro spheres with embedded gold detected with secondary electrons (left) and back scattered electrons (right)

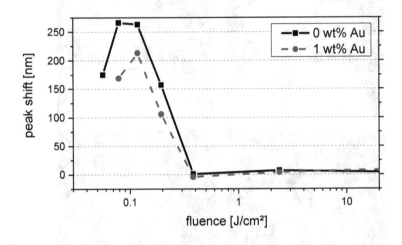

Figure 48: Peak shift plotted versus laser fluence for PLML of pure ZnO and ZnO with
1wt% supported Au NP after 50 passages laser irradiation with 40 ns pulse
length, 355 nm wavelength and 86 kHz repetition rate

The possibilities of PLML generating hybrid materials is wide. Here, only proof-of-principle experiments showing fundamental possibility are described, whereas future experiments should consider the nanoparticle size, the amount of nanoparticles and fluence. Furthermore, it might be particularly interesting to explore whether nanoparticles that do not absorb on microparticles surface can be embedded by PLML. Considering the mechanism reported by Ishikawa et al. [Ishikawa2010], variation of fluence could play an important role in understanding the change of chemical composition of particles by PLML. In the future, using copper nitride particles and exposing them in EA (as investigated in chapter 4.1.5.1) to fluences causing particle melting would be interesting to study whether metallic copper SMS can be obtained. Additionally, educt nanoparticle size and final nanoparticle size embedded into SMS matrix should be investigated.

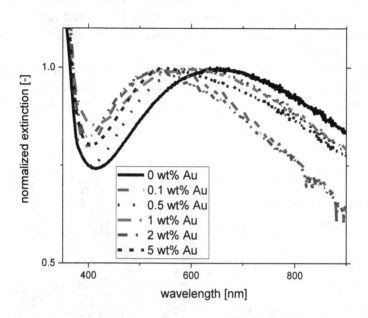

Figure 49: UV-vis spectra after 50 passages for different amount of gold nanoparticles for optimized PLML process parameters

4.2.2 Inclusion of supported gold nanoparticles into their semiconductor support

Synopsis

Following the first proof-of-principle in chapter 4.2.1 showing how ZnO particles can be converted into ZnO SMS and that gold nanoparticles can be integrated into the ZnO SMS if supported on the educt semiconductor a more detailed study concerning the material, laser and process parameter is required.

Hence, the controlled formation of ZnO SMS with integrated crystalline Au NP is studied and demonstrated in the following chapter. Due to sequential irradiation, using the passage reactor design allows identifying intermediates formed during the conversion of monodisperse gold nanoparticles supported on ZnO to integrated Au NP into ZnO SMS, as schematically illustrated in Figure 50.

This chapter addresses exemplary studies in the integration of supported nanoparticles into their solid support namely gold nanoparticles into zinc oxide sub-micrometer spheres by energy-controlled pulsed laser melting in a free liquid jet.

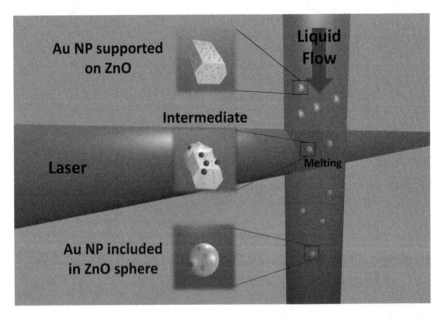

Figure 50: Schematic illustration of the exemplary studies on the integration of supported nanoparticles into their solid support, namely gold nanoparticles into zinc oxide sub-micrometer spheres, by controlled pulsed laser melting in a free liquid jet [Lau2015]

M. Lau, A. Ziefuss, T. Komossa, S. Barcikowski, *Inclusion of supported gold nanoparticles into their semiconductor support*, Phys. Chem. Chem. Phys. 17 (**2015**) 29311-29318

Abstract

Supported particles are easily accessible as standard materials used in heterogeneous catalysis and photocatalysis. This article addresses our exemplary studies on the integration of supported nanoparticles into their solid support, namely gold nanoparticles into zinc oxide sub-micrometer spheres by energy controlled pulsed laser melting in a free liquid jet. This one-step, continuous flow-through processing route reverses the educt's structure, converting the ligand-free surface adsorbate into a spherical subsurface solid inclusion within its former support. The results show how a nanoparticulate surface adsorbate can be included in form of crystalline nanoparticles into the resolidified support matrix demonstrated by using plasmonic nanoparticles and semiconductor microparticles as reference materials.

Introduction

Supported metal nanoparticles on metal oxides are commonly available raw materials with application prospects in catalysis [Stratakis2012], [Yu2012], [Meier2014]. For example, gold nanoparticles (Au NP) on zinc oxide (ZnO) support catalyze chemical reactions [Strunk2009]. Even though gold nanoparticles (Au NP) can be easily attached to ZnO nanoparticles [Li2011], [Lau2014e] their integration into a zinc oxide matrix on the nanoscale is challenging. A possibility to melt aqueous dispersions of particulate materials under non-equilibrium conditions is the pulsed laser melting in liquid (PLML). PLML has been shown to be an effective synthesis route for crystalline sub-micrometer spheres (SMS), but additive PLML has only rarely been investigated and this has not covered supported particles yet.

The fabrication of SMS by PLML is a technique pioneered by Koshizaki and coworkers [Ishikawa2007], [Wang2011]. Early application examples have been demonstrated by Hu et al. for SMS in lubricant oils to reduce the friction coefficient [Hu2012], and Fujiwara et al. who used ZnO SMS as a light emitting laser source [Fujiwara2013].

Since the first report several articles report the possibility and power of this technique. In contrast to size-reducing pulsed laser fragmentation in liquid (PLFL), lower laser fluences are required, resulting in particle melting and resolidification as spheres. Wang et al. found the onset laser fluence for particle melting of zinc oxide to be between 33 mJ/cm² and 67 mJ/cm² [Wang2011]. A limitation to high laser fluences was reported by Wang et al. for the laser melting of CuO nanoparticles to copper SMS [Wang2012]. At a fluence of 150 mJ/cm² the SMS showed a rough surface. This indicates a less effective particle melting due to onset of particle fragmentation. The influence of educt particle size on melting effectivity was demonstrated by Tsuji et al. [Tsuji2013a]. They showed that by increasing the educt particle size of zinc oxide particles the fluence required for sufficient PLML to obtain monomodal and smooth ZnO SMS increases. Namely 70 nm ZnO educt particles were molten completely at fluences of 100 mJ/cm² whereas aggregated (500 nm aggregate diameter) ZnO educt particles were only partially molten at laser fluences of 200 mJ/cm² and completely molten at fluences of 300 mJ/cm² [Tsuji2013a]. Liu et al. demonstrated that PLML appears to be an isochoric process using octaedrical Au NPs, and transformed them into monodisperse Au nanospehres preserving the volume of the particles, thus demonstrating that PLML can be applied as isochoric particle reshaping method [Liu2015].

First reports on the laser melting of Au NPs were given by El-Sayed and coworkers [Link1999a], [Link1999b], [Link1999c] who reported the sufficient interaction of surface plasmons with 532 nm laser light for the isochoric melting of Au nanorods to nanospheres. In 2005 Inasawa et al. showed that for gold nanoparticles the melting temperature is approximately 100 K below the value of the bulk material using nanoparticles around 38 nm and applying a laser wavelength of 355 nm [Inasawa2005]. Tsuji and coworkers recently reported the possibilities to melt aggregated Au NPs to SMS with 532 nm laser light and found a PLML induction time interval caused by the laser-induced surfactant removal [Tsuji2013], [Tsuji2015].

Studies on SMS formation mostly report the use of 355 nm laser wavelength and nanosecond pulses to cause distinctive particle melting in liquid, even though other wavelengths and shorter (ps) pulse lengths are able to cause (partial) melting as well [Lau2104a].

Besides simple physical melting, changes in chemical composition are also reported [Ishikawa2007], [Wang2012], [Ishikawa2010]. Ishikawa et al. studied the formation of boron carbide SMS from boron nanoparticles in organic solvents. Additionally, PLML-alloying of non-equilibrium phases is known

[Swiatkowska-Warkocka2013]. Swiatkowska-Warkocka et al. generated bimetallic crystalline SMS of copper and gold, but they obtained a solid solution in contrast to the inclusion type of SMS reported here. Nakamura et al. recently reported on magnetite particles integrated into calcium phosphate SMS by the combination of chemical precipitation and a laser wavelength absorbance of the iron salt, which was subsequently integrated into the calcium salt SMS during PLML [Nakamura2015]. Hence, the investigation on the integration of iron into SMS included chemical reactions and complex precipitation-reaction-melting interdependence [Nakamura2015]. Next to reactive PLML or the alloying of miscible elements, the question arises as to what happens during the irradiation of partly miscible or immiscible mixtures under PLML conditions. PLML of a colloidal mixture of gold with as-prepared iron oxide nanoparticles resulted in a core-shell structure with a porous surface after Fe etching [Kawaguchi2006], [Kawaguchi2007], [Swiatkowska-Warkocka2012], but supported particles have not yet been investigated.

In this context the facile synthesis of supported particles comes into play, such as the adsorption of ligand-free metal nanoparticles (Au, Ag, Pt, Pd) on supports (ZnO, TiO_2, $BaSO_4$, graphene) [Wagener2012a] simply by mixing at defined liquid parameters [Marzun2015], thus achieving up to a 60 wt% NP-on-support at 100% yield. Taking Au/ZnO as a relevant example (e.g. for photocatalysis applications), it was uncertain what happens if this nano/micro support system is exposed to PLML and motivating the present study.

Experimental

Preparation of supported particles follows mechanism firstly reported by Wagener et al. [Wagener2012a] and described in detail by Marzun et al. [Marzun2015], where Au/ZnO preparation procedure was analogous to Lau et al. [Lau2014e]. Au NP were generated by picosecond pulsed laser ablation in liquids (PLAL) in 600 µM aqueous phosphate buffer to receive small particles and size separation was conducted by centrifugation using an Ultracentrifuge (Beckman Coulter), with a force of 30,000 g for 14 minutes resulting in a monodisperse Au NP colloid with defined concentratio. Concentration of supernatant was determined from extinction at a fixed wavelength subsequent to calibration with different concentration of same particle size. The monodisperse Au colloid had a concentration of 62.7±0.1 µg/mL after centrifugation. We added 95 mg zinc oxide particles to an overall colloid volume of 450 mL Au NP in water resulting in an Au NP loading of 30 wt% referred to zinc oxide after 100 % Au NP adsorption.

Detailed particle size distributions (mass-weighted 5±0.9 nm, number-weighted 4.7±0.8 nm) and determination of concentration by calibrated UV-vis spectroscopy of the monodisperse Au NP colloids (polydispersity index < 0.03) are provided in Fig. 8-38 of supporting information. Zinc oxide particles (Sigma Aldrich) were simply added to the monodisperse gold nanoparticle suspensions and the obtained gold/zinc oxide supported particle powder was dried (50°C, 8h). Note that there are different values describing the amount of supported nanoparticles. Fig. 8-39 illustrates the absolute wt% of nanoparticles plotted versus the wt% of gold nanoparticles adsorbed onto zinc oxide as support, resulting in nonlinear correlation between the wt% and vol%. The volume percentage might be of interest as optical response and colloidal analytics often correlate to the particle volume. But in catalysis application wt% is the standard unit. For PLML the highest amount of supported gold nanoparticles used in this study was 30 wt% of gold adsorbed on the support (thus a gold amount of approx. 23 wt% absolute), equivalent to 11 vol% (see Fig. 8-39).

Laser irradiation was performed with an Nd:YAG laser operating at the 3rd harmonic (355 nm wavelength) with 85 kHz repetition rate, 23 watt and a pulse length of 40 ns.

For determination of the influence of laser fluence on PLML and defined volume-specific laser energy dose we used the design of a sequential liquid flow. This design has been reported previously and allows to study the impact of the applied laser fluence while focusing into a thin liquid filament [Lau2014a]. Hence, minimized beam path in liquid significantly reduces fluence variation along beam propagation, allowing defined energy balancing. As reported before, a strong shift of a local peak in the UV/vis spectrum can be attributed to sufficient PLML effect [Lau2014a]. Thus, plotting the observed shift of this local peak can be used to draw conclusion for PLML efficiency and the applied laser fluence. To study the influence of the applied laser fluence pure zinc oxide particles were used.

PLML of the supported Au/ZnO was conducted under optimized conditions found for pure zinc oxide in pure water with no additives. All used suspensions for laser irradiation had a particle concentration of 0.1 wt% as this was found to be appropriate to characterize PLFL and PLML sufficiently [Lau2014a].

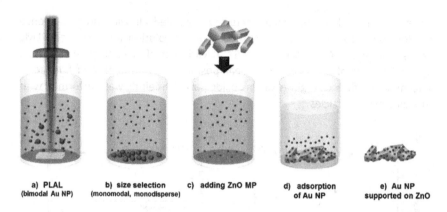

a) PLAL
(bimodal Au NP)

b) size selection
(monomodal, monodisperse)

c) adding ZnO MP

d) adsorption
of Au NP

e) Au NP
supported on ZnO

Figure 51: Schematic illustration of the process steps to fabricate supported Au NP/ZnO MP with monodisperse nanoparticles attached to the support: a) picosecond-pulsed laser ablation in water (PLAL); b) size selection by centrifugation; c) addition of zinc oxide microparticles to the gold colloid; d) adsorption of gold nanoparticles onto the zinc oxide microparticles; e) dried powder for further PLML processing

Size characterization of the particles was performed with an analytical disc centrifuge (CPS instruments) at 24,000 rpm, extinction of the colloids was determined with an UV-vis absorbance spectrometer (evolution 201) in a quartz glass cuvette. Diffuse powder reflection was determined in a spectrometer (Varian Cary) using a spectralon reference (PTFE). SEM images are taken with an SEM (FEI Quanta 400) on carbon supports. XRD measurements of the compound particles before and after PLML were carried out with a Cu K-α irradiation source at 40 kV and 40 mA in a Bruker D8 Advance system.

The preparation route for the nano/micro support systems used here is schematically shown in Fig. 51.

Results and Discussion

We demonstrate how plasmonic nanoparticles can be embedded into semiconductor sub-micrometer spheres by laser melting in continuous liquid flow. This route allows achieving an integration of metallic nanoparticles into a semiconductor SMS matrix in one step without performing any chemical precursor-based synthesis. Within our experiments only pure water was used as

carrier liquid.   A fluid jet reactor setup was applied allowing precise fluence
control and sequential analysis of the product evolution after defined PLML
passage numbers. As shown in Fig. 52 a) variation of laser fluence results in
significant shifts of the local UV-vis peak positions. Variation of laser fluence to
determine the fluence regime for zinc oxide particle melting was performed with
pure zinc oxide particles.

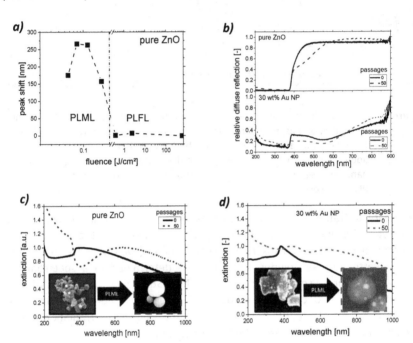

Figure 52:   a) Impact of laser fluence on UV-vis peak shift of zinc oxide suspension after
laser irradiation (at 355 nm and 40 ns); b) relative diffuse scattering of the
dried particle powders before and after 50 passages PLML for pure ZnO (top)
and 30 wt% Au NP supported on ZnO (bottom), respectively; c) normalized
(at peak in visible regime) UV-vis spectrum of pure zinc oxide particles before
(black/solid line) and after (blue/dashed line) PLML in pure water with SEM
images of corresponding educt and product particles as inset; d) normalized
UV-vis spectrum of zinc oxide particles with 30 wt% supported gold
nanoparticles before (black/solid line) and after (red/dashed line) PLML in
pure water

Corresponding UV-vis spectra for pure zinc oxide from which peak shift is determined, are shown in Fig. 8-40 in the supporting information. The bathochromic shift of the peak can be correlated to ZnO particle melting and resolidification as spheres.

Thus plotting this shift versus the laser fluence illustrates the process window for PLML (at 355 nm laser wavelength and 40 ns pulse length) indicated in Fig. 52 a) to be ≤ 0.2 J/cm². Fig. 52 b) shows powder scattering spectra for pure zinc oxide and for the supported Au NP on ZnO MP with 30 wt% gold loading. The diffuse reflection spectra show that increased extinction at wavelengths above 600 nm results from light scattering [Stamatakis1990]. This diffuse reflection is caused by spheres formed by PLML as proposed by Wang et al. [Wang2011]. That such spheres reflect light wavelengths in their size regime is confirmed here by scattering measurement of the dried particle powders. Additionally, absorption of Au NP Relative minimum in the scattering spectra around 550 nm) can be observed from diffuse reflection which becomes more distinctive after 50 passages PLML due to size increase of Au NP. UV-vis spectra before and after 50 passages of laser melting in liquid filament are shown in Fig. 52 c) and d) with SEM images of the corresponding educt and product particles as insert. For the semiconductor SMS products with gold nanoparticle inclusion two peaks in the visible regime are obtained. For the educt gold nanoparticle size a plasmon peak around 550 nm is observed. In the course of laser melting and reversing particle structure from supported Au NP to solid Au NP inclusions, obviously Au NP size increases as validated by SEM pictures (Fig. 8-42).

UV-vis spectra in Fig. 53 show the difference between PLML of pure ZnO and for PLML of 30 wt% Au NP with ZnO. Evolution of the local peak around 600 nm can be attributed to scattering of light on formed spheres with sizes in this wavelength regime. This scattering effect occurs for pure ZnO SMS and for ZnO SMS with included Au NP. An increase of UV extinction for pure ZnO is also observed during PLFL due to formation of defect-rich ZnO particles [Lau2014a]. The ZnO SMS obtained after PLML appear yellowish, similar to defect-rich, bandgap-shifted ZnO particles derived from PLFL of ZnO [Lau2014a].

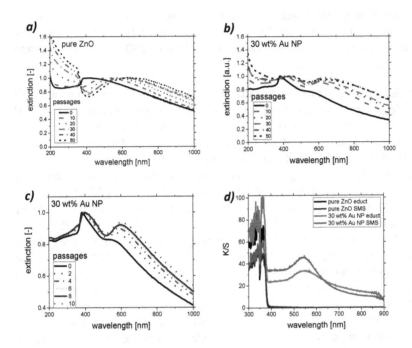

Figure 53: Evolution of UV-vis spectra during PLML with a fluence of 80 mJ/cm²
normalized at a peak in the visible regime for a) pure zinc oxide after each 10
passages from 0 to 50 irradiation cycles, for b) 30 wt% Au NP on ZnO after
each 10 passages from 0 to 50 irradiation cycles and c) for 30 wt% Au NP on
ZnO after every second cycle during first 10 passages; d) ratio of absorption
to scattering K/S determined from Kubekla-Munk theory for pure ZnO before
and after 50 passages PLML and for 30 wt% Au NP on ZnO before and after
50 passages PLML, respectively

Thus we attribute the increased UV extinction to the formation of defect-rich ZnO
SMS what is supported by fluorescence measurements shown in Fig. 8-43. For
inclusion of Au NP into ZnO SMS this distinctive increase of UV extinction is not
observed. Instead a second extinction peak around 400 nm to 430 nm occurs.
Based on the Kubelka-Munk equations [Kubelka1931], [Kubelka1948] we
determined the ratio of absorption to scattering K/S from the diffuse reflection
spectra, shown in Fig. 53 d). Absorption caused by plasmon resonance of Au NP
around 540 nm can be observed for educt particles as well as for Au NP inclusions
in ZnO SMS. The plasmon peak of the Au/ZnO educt is boarder and less distinctive
due to smaller Au NP on ZnOs surface (see Fig. 8-44 and 8-45, 0 passages). After

50 passages of PLML this Au NP is more distinctive due to size increase of gold. From the diagram in Fig. 53 d) we determined the bandgap of the different educts and products. Diagrams and linear fits for determination of band gap energy, based on Kubelka-Munk equation and absolute values of diffuse reflection spectra are shown in Fig. 8-46 in supporting information. Determination of bandgap energy shows that for both pure ZnO before and after 50 passages of PLML as well as for untreated 30 wt% Au NP on ZnO the bandgap energy is around 3.23 eV (~384 nm). When Au NP are included into ZnOs matrix this value decreases to around 3.07 eV (~404 nm).

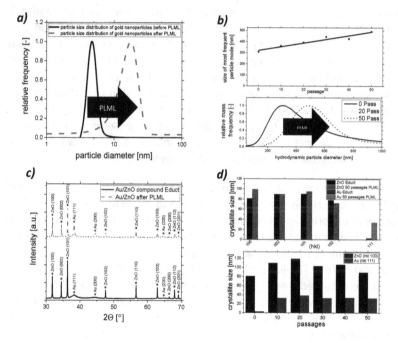

Figure 54: a) Particle size distribution of gold nanoparticles before (red/dashed curve, supported on zinc oxide) and after (black/solid) curve, integrated into zinc oxide sub-micrometer spheres) PLML detected by analytical disc centrifuge measurements (before PLML) and determined by a Gaussian fit from histograms taken from SEM images (after PLML) for 30 wt%; b) size shift and particle size distribution of the 30 wt% Au NP on ZnO compound after 0, 20 and 50 passages; c) XRD pattern of the gold zinc oxide compound with 30 wt% Au NP before and after 50 passages exposed to PLML conditions and d) cristallite size calculated from the XRD pattern and evolution of crystallite size after each 10 passages from 0 to 50 irradiation cycles

This is in agreement witch Chanu et al. who integrated Au clusters into ZnO [Chanu2014]. In similar, the preservation of plasmon resonance for integrated plasmonic silver nanoparticles into TiO2 and related photocatalytic activity was demonstrated by Awazu et al. [Awazu2008]. The Au NP integrated into ZnO SMS also possess plasmonic properties shown in Fig. 53 d).

Obviously, during PLML of supported Au NP on ZnO a significant increase of Au NP size is observed at almost constant ZnO volume. Fig. 54 a) shows the particle size distribution of gold nanoparticles before and after PLML. The size of the Au NP measured by SEM (see Fig. 8-41) increases after PLML from ~5 nm to ~17 nm, by about a factor of 3. This means that zinc oxide microparticles and gold nanoparticles obviously have been transferred into a molten state and both resolidify as spheres with gold granules inside ZnO. The evolution of crystallite size shown in Fig. 54 d) proves that already after 10 passages Au NP with diameters around 30-40 nm are formed and do not chance significantly during the additional passages. Increase of Au NP size causes shift of surface plasmon resonance wavelength and increase of intensity. Thus embedding plasmonic particles into semiconductors causes enhancement of light absorption, becoming naturally more distinct if size of Au NP is increased, as known for silver nanoparticles integrated into semiconductors [Spinelli2012].

XRD pattern reveal crystallinity of the two materials whereby a size increase of the gold nanoparticles is confirmed (Fig. 54 c) and d)). SEM images in Fig. 55 c) and d) validate fabrication of sub-micrometer spheres for zinc oxide with 5 wt% Au NP and 30 wt% Au NP at a laser fluence of ~ 80 mJ/cm² at liquid jets' surface. This is in agreement to Tsuji et al. who observed similar SMS particle sizes after PLML at 100 mJ/cm² [Tsuji2013a]. At laser fluences of 380 mJ/cm² or higher no PLML can be observed (see Fig. 52 a) and Fig. 8-40) tested for pure ZnO. But a reduction in hydrodynamic particle diameter occurs (starting at laser fluences above ~ 200 mJ/cm²), confirmed by ADC analysis of hydrodynamic particle diameter shown in Fig. 8-42. For lower laser fluences no or only slight increase in particle size is obtained, indicating an almost isochoric particle melting (of primary particles and aggregates) and resolidification, similar to the findings on particle reshaping by Nakamura et al. and Liu et al. [Liu2015], [Nakamura2015].

Integration of the Au NP into the volume of its ZnO support was determined by correlated SEM images taken both with a secondary electron detector and a back scattered electron detector (Fig. 55 c), d)).

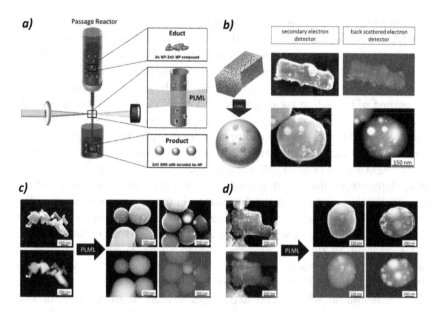

Figure 55: a) Illustration of the set-up for laser irradiation of particles in a free liquid flow changing particles morphology from educt to product downstream during PLML; b) schematic illustration of the Au NP/ZnO MP compound before PLML with nanoparticles onto the support and after PLML with inclusion of nanoparticles (left) and SEM images of corresponding particles (right); c) zinc oxide particles with 5 wt% of gold nanoparticles before (left) and after (right) PLML (50 passages) imaged with a secondary electron detector (top row) and a back scattered electron detector (bottom row); d) zinc oxide particles with 30 wt% of gold nanoparticles before (left) and after (right) PLML (50 passages) imaged with a secondary electron detector (top row) and a back scattered electron detector (bottom row)

The former is more sensitive to surface topography (lower penetration depth of electrons) and the latter is more sensitive to elemental contrast (higher penetration depth), thus depicting the gold nanoparticles in higher contrast to zinc oxide. Additional SEM images are shown in Fig. 8-44 and 8-45. A quantitative transformation of surface-adsorbed gold into inclusions inside zinc oxide sub-micrometer spheres is obtained after 50 passages of laser irradiation with 80 mJ/cm². SEM images showing the evolution of Au NP inclusion into ZnO SMS after each 10 passages are depicted in Fig. 8-47.

Regarding the control of the educt mass flow, Figure 55 a) illustrates the process of PLML in a sequential liquid flow with defined volume flow rates. The liquid jet is formed by a capillary and the laser is focused perpendicular on the liquid. Particles passing the irradiated volume are transferred into the molten state. Fig. 55 d) shows that different sizes of included Au NP can be obtained. Due to cumulation of irradiation cycles (passages) it might be possible that previously formed SMS are remolten thus giving the possibility of Au NP fusing to larger NP. Figure 55 b) sketches the transformation from supported Au NP onto ZnO towards Au NP inclusions within ZnO SMS by PLML, with corresponding SEM images. SEM images before PLML and after 10, 30 and 50 passages, shown in Fig. 56 a), demonstrate that Au NP size increases already after 10 passages and are not completely included into the ZnO matrix in the intermediate state of the processing. This formation of crystalline Au NP is in agreement with sizes determined by XRD peak analysis, shown in Fig. 54 d). Scherrer equation reveals that crystalline Au NP with diameters around 30 nm from already after 10 passages and do not change their size significantly afterwards. The SEM images after 10 passages (Fig. 56 a)) demonstrate different appearance of non-included and included Au NP. After 30 and 50 passages no more Au NP absorbed on ZnO particles surface are observed, but their inclusion into the ZnO spheres is obvious. A mechanistic hypothesis of process evolution is illustrated in Fig. 56 b).

From UV-vis spectra it can be concluded that generated Au/ZnO hybrid SMS feature interesting optical properties with two local extinction peaks as known for gold nanorods but with completely spherical particles.

Conclusions

In summary, integration of metals, in particular plasmonic nanoparticles into a submicron-confined semiconductor (or dielectric) volume is challenging but could provide unique structural and optical properties. Here it is shown that pulsed laser melting of supported gold/zinc oxide particles enables embedding of plasmonic nanoparticles into a semiconductor matrix isochorically forming monomodal sub-micrometer spheres. The use of monodisperse and ligand-free gold nanoparticles allowed high nanoparticle loads onto zinc oxides surface and to investigate its structural inversion. During processing gold nanoparticles first increase in size and are subsequently transferred during additional irradiation passages into their spherical support. The investigation of structural morphology and the elemental contrast of the particles by correlated electron microscopy confirmed the integration of the metal nanoparticle into the semiconductor solid spheres. The liquid flow passage reactor allows to characterize the material properties after each passage with a defined laser energy dose [Lau2014a], providing sequential "snapshots" on the evolution of the process and mechanistic insight into pulsed laser melting of supported particles. Hence nanoparticulate surface adsorbates, widely available as raw materials for catalysis, have been integrated as solid inclusion into their support.

ACKNOWLEDGMENTS

We thank Bernardo Oliviera de Viestel for experimental support during PLML of the gold/zinc oxide hybrid particles.

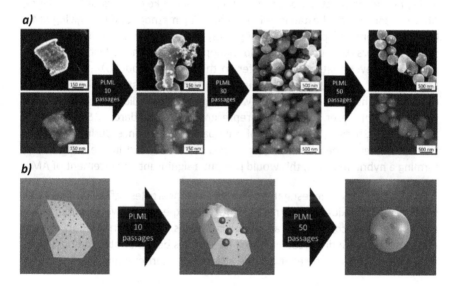

Figure 56: Evolution of crystalline Au NP inclusion into their ZnO support for different number of irradiation passages; a) SEM images of educt (left) and after 10, 30 and 50 passages of PLML detected witch a secondary electron detector (top images) and a back scattered electron detector (bottom images) showing evolution of Au (bright) and ZnO particle morphology b) schematic illustration of the evolution from the adsorbed Au NP on ZnO (educt) via intermediate state with size-increased Au NP (see XRD in Fig. 54) on reshaped ZnO towards included Au NP into the spherical ZnO matrix (product)

4.2.3 Near-field-enhanced, off-resonant laser sintering of semiconductor particles for additive manufacturing of dispersed Au-ZnO-micro/nano hybrid structures

Synopsis

The most important additive manufacturing (AM) technologies are selective laser melting (SLM), selective laser sintering (SLS) and laser metal deposition (LMD) [Gu2012]. These generative manufacturing processes allow fabricating unique parts from a powder material [Murr2012], whereby integration of functionality and light weight constructions can be named as the key factors differentiating AM from conventional-established manufacturing methods and thus giving AM a huge and promising potential [Huang2013]. Here, a way of amplifying SLS of semiconductors by supporting gold nanoparticle on the educt particles is presented. Several later reports refer to this study [Schade2014, Garcia2014, Marzun2014, Gökce2015, Manshina2015]. In contrast to the previous experiments, here the fundamental wavelength of a cw solid state Nd:YAG laser was used for laser processing representing a standard SLS procedure [Kruth2003]. Studying the impact of nanoparticles' presence during AM holds particular interest because if nanoparticles are embedded in the solid matrix forming a hybrid material, this would present a significant advancement of AM.

This research article investigates how gold nanoparticles influence the laser sintering of zinc oxide microparticles. Note that experiments were conducted in ambient gas, in contrast to all previous studies of laser irradiation, where nanoparticles were dispersed in water. Hence, this demonstrates the possibility of supported particle processing in the absence of a carrier liquid.

M. Lau, R. Niemann, M. Bartsch, W. O'Neill, S. Barcikowski, *Near-field-enhanced, off-resonant laser sintering of semiconductor particles for additive manufacturing of dispersed Au–ZnO-micro/nano hybrid structures*, Appl. Phys. A 114 (**2014**) 1023-1030

Abstract

Off-resonant near-field enhancement by gold nanoparticles adsorbed on crystalline zinc oxide significantly increases the energy efficiency of infrared laser sintering. In detail, ten different gold mass loads on zinc oxide were exposed to 1,064 nm cw-laser radiation. Variation of scan speed, laser power and spot size showed that the energy threshold required for sintering decreases and sintering process window widens compared to laser sintering of pure zinc oxide powder. Transmission electron microscope analysis after focused ion beam cross sectioning of the sintered particles reveal that supported gold nanoparticles homogenously resolidify in the sintered semiconductor matrix. The enhanced sintering process with ligand-free gold nanoparticles gives access to metal-semiconductor hybrid materials with potential application in light harvesting or energy conversion.

Introduction

Additive manufacturing by selective laser sintering produces 2D and 3D geometries with unique structures made of metals, insulators, or ceramics [Hong2013, Agarwala1995, Tolochko2000, Dudziak2010].

Nevertheless, if the process is applied for infrared (IR) laser-transparent materials like [Zhu2007, Kathuria1999] semiconductors, manufacturing of parts usually is inefficient due to low incoupling of laser energy. To relieve inefficiency either the laser wavelength can be changed toward the band gap energy of semiconductors, or the material response to the IR wavelength can be enhanced. The former is to be understood as resonant and the second off-resonant laser incoupling for sintering. Resonant laser sintering has been shown by Crespo-Monteiro et al. who examined the influence of plasmonic silver nanoparticles (NP) to the required laser intensity to cause sintering [Crespo-Monteiro2012]. They showed that silver nanoparticles added to a mesoporous TiO_2 film enabled sintering at lower intensities with 488 nm laser light, hence exciting resonant surface plasmons of the added nanoparticles close to their plasmon band after adsorption (488 nm) [Crespo-Monteiro2012]. Alternatively, also off-resonant excitation of gold particles by infrared laser radiation has been reported, making use of the near-field enhancement around the NP [Boulais2012]. Hence, this method relies on the scattered near-field instead of energy absorption in the particle itself, in particular for ultra-short-pulsed lasers. For laser sintering cw lasers are widely applied, so that off-resonant near-field amplification could be a strategy to incouple energy into otherwise laser-transparent materials by supporting gold nanoparticles.

For this the amount of plasmonic particles should play an important role in the sintering process. For the investigation of off-resonant laser sintering process window we changed material response to laser irradiation by varying the amount of surface-adsorbed gold nanoparticles on zinc oxide microparticles. Resonant laser sintering operates with shorter laser wavelength bearing the risk to approach ablation regimes by interband absorption, while off-resonant laser sintering adapts material parameters to IR laser light.

Pure zinc oxide is of interest as semiconductor with a direct band gap around 3.3 eV at 300 K and large exciton binding energy (60 meV) [Özgür2005]. Further it is a cheap and an easy accessible material compared to rare earths, and recent efforts focus on, e.g. aluminum doped zinc oxide as transparent conductive thin films to substitute expensive indium tin oxide ITO [Bai2006].

Gold nanoparticles adsorbed on metal oxides are widely used in heterogeneous catalysis [Haruta1997, Hashmi2006] and ZnO-supported gold nanoparticles (ZnO@Au) were reported to enhance conversion efficiency of dye-sensitized solar cells [Dhas2008] and to improve photocatalytic activity for high performance lithium ion batteries [Ahmad2011].

We present a strategy to enhance infrared laser sintering by gold nanoparticles (Au-NP) adsorbed on zinc oxide micro particles (ZnO-MP). For this purpose, ligand-free colloidal Au-NP generated by pulsed laser ablation in liquids (PLAL) are an adequate candidate as they absorb with almost 100% efficiency on micro particle surfaces [Barcikowski2013, Wagener2012a]. PLAL technique was first reported by Henglein and Fojtik [Fojtik1993] and since then drawn a steadily grown interest, as Au-NP in pure water show a remarkable high electrostatic stability without any surfactants [Barcikowski2009, Rehbock2013]. Here we combine the techniques of PLAL and laser sintering to enhance sintering processes and at the same time fabricating a nano/micro metal/semiconductor hybrid material.

Experimental

Preparation of gold nanoparticles and adsorption on zinc oxide microparticles

Gold nanoparticles were produced by pulsed laser ablation in liquid (PLAL) with a picosecond-pulsed Yb:YAG laser (1064 nm) at 75 µJ pulse energy and 100 kHz focused on a gold target (99.00 % purity, Goodfellow) placed in an ablation chamber designed for PLAL filled with 100 ml deionized water [Nachev2012, Wagener2010]. The concentration of obtained gold colloid was determined by differential weighting of the gold target on a micro balance. After preparation of the Au-NP different amounts of ZnO-MP were suspended in the gold colloid to achieve corresponding loadings from 0.001 wt% to 0.5 wt% gold on ZnO. The colloidal Au-NP adsorbed completely and homogenous onto micro particle support within several hours. By subsequent sedimentation of the microparticles the transparent supernatant was decanted and particles were dried at 323 K for 12 hours. After de-compaction in a mortar ZnO-MP@Au-NP powders with weight ratios of 0 wt%, 0.001 wt%, 0.005 wt%, 0.01 wt%, 0.05 wt%, 0.1 wt%, 0.5 wt%, 1 wt%, 2 wt%, and 5 wt% shown in Figure 57 were received.

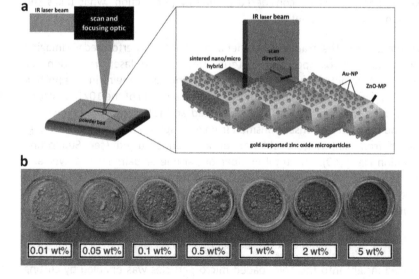

Figure 57: a Illustration of laser sintering process and enhancement via gold nanoparticles supported on zinc oxide micro particles. b Zinc oxide micro particle powder with increasing weight percentage of gold nanoparticles adsorbed on the particle surface.

Laser sintering of Au-NP supported ZnO-MP

To cover a broad spectrum of material response during laser sintering from no reaction to strong ablation and to obtain the sintering process in appropriate manner, parameter studies of varying laser power and writing speed with two different spot sizes at the surface of powder bed were conducted. For this purpose a powder bed of semiconductor micro particles has been prepared by manually compacting particles with a stamp, to obtain an even surface. Laser sintering has been performed with a fiber laser operating at 1064 nm in cw-mode (G3, SPI Lasers PLC, redENERGY, SM-series). Applied laser power varied from 2 W to 20 W in steps of 2 W. Writing speed of the laser has been varied from 10 mm/s to 5000 mm/s. Each powder type with different gold load was treated with a 10x10 matrix with corresponding parameter pairs of power and writing speed. In order to have a detailed view of sintering behavior the laser beam was slightly focused to a spot size of 150 µm and for a stronger material response same parameters were applied with a spot size of 31 µm, resulting in higher intensities (up to 113,2 kW/cm² and 2,653 MW/cm², respectively). In the following the applied intensities will be considered as energy per section, what takes the irradiation time into account.

Characterization of the irradiated powder bed surface was performed via imaging by an optical microscope. The material response to laser irradiation for mentioned parameters was evaluated from images shown in Supporting Information (Fig. 8-48) taken by the microscope. Each field of the 10x10 matrices from 10 different gold loadings was evaluated and rated into the following 5 classes: no material response, transition from no response to sintering, sintering, transition from sintering to ablation or ablation occurred (see Supporting Information Fig. 8-49). The total number of parameter pairs laser power and writing speed (for 150 µm spot diameter) resulting in successful sintering gives a quantitative value of the process window area. In the shown power-speed diagrams this area is marked green.

FIB cutting of supported micro particles and TEM analysis

A detailed insight into the gold-loaded micro particles was enabled by cutting particles by a focused ion beam (FIB) and inspection of particle cross section in the transmission electron microscope (TEM). SEM pictures taken during the FIB cutting process are shown in Figure 8-48.

Results and Discussion

The conducted experiments with varying power and writing speed show a curtail influence of gold nanoparticles to material response on laser irradiation. At high gold loadings on zinc oxide microparticles the process window significantly extends compared to pure ZnO-MP. Pictures taken with the optical microscope and the corresponding power-speed diagrams are shown in Figure 58 as well as the corresponding laser energy per section. The material response of pure ZnO-MP differs compared to 5wt% Au-NP loaded ZnO-MP regarding the threshold energy required for sintering. For 5 wt% gold nanoparticles on ZnO-MP sintering threshold decreases by a factor of 2.3 down to around 1 J/m. At this it is impossible to define intensity thresholds or energy per section thresholds independently, as the former is not considering residence time of irradiation and the latter gives no information about focusing conditions. From diagrams shown in Figure 58 and from Supporting Information it can be seen that higher intensities tend to lead to lower residence time to cause sintering, therefore reducing required energy per section.

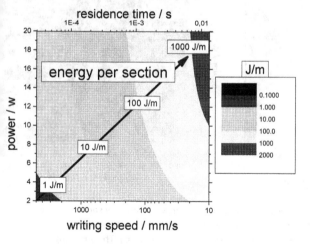

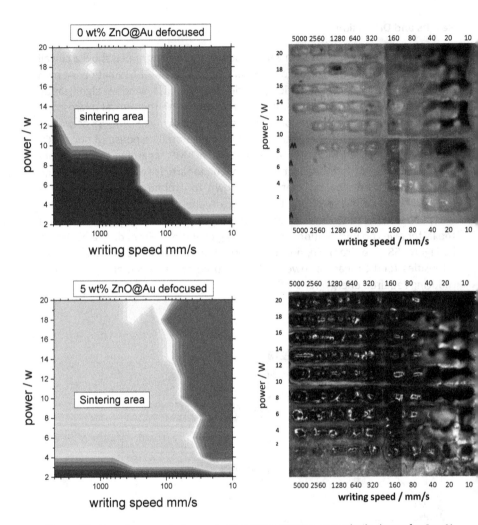

Figure 58: Top: Process window and microscope pictures exemplarily shown for 0 wt% and 5 wt% of gold nanoparticle loading on zinc oxide (see Supporting Information Figure S2 for all process window diagrams). The sintering process window (marked green) has been quantified for all different gold loadings and plotted in figure 3. Bottom: energy per section resulting from the applied power and writing velocities.

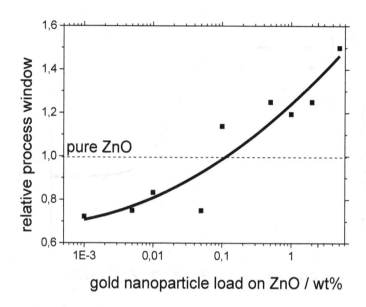

Figure 59: Process window for laser sintering in dependence of gold nanoparticle load on zinc oxide microparticles (parameters of laser power, writing speed and laser spot size given in supporting information Fig. 8-49).

Figure 59 shows the effect of gold nanoparticles on zinc oxide microparticles for the sintering process window. All values of the process window area are standardized to the value for pure zinc oxide. Low loadings of gold (0.001, 0.005, 0.01 and 0.05 wt%) result in narrower sintering process window, whereas loadings of 0.1 wt% significantly widens sintering process window up to 50%.

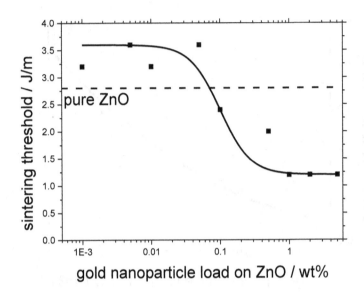

Figure 60: Sintering threshold (minimum required energy per section) for writing
speeds of 5 m/s for 150 µm spot size. Black dashed line shows threshold for
pure zinc oxide microparticles without gold. Necessary energy per section
slightly increases for low gold nanoparticle loadings but further decreases.
Findings for sintering thresholds correlate with the widening of sintering
process window, shown in figure 58.

Figure 60 depicts the sintering thresholds for different gold loadings obtained at
5 m/s with 150 µm spot size. The findings correlate with influence of gold
nanoparticles to sintering process windows. For low gold loadings (up to 0.05
wt%), minimum required energy per section to cause sintering slightly increases
from 2.8 J/m for pure zinc oxide microparticles to around 3.6 J/m. If amount of
gold nanoparticles is 0.1 wt% or higher, energy per section necessary to cause
material response is reduced. Sintering threshold decreases down to 1.2 J/m for
1 wt%, 2wt% and 5 wt% of gold nanoparticles attached to microparticle support.
Therefore the highest reduction of sintering threshold by surface-adsorbed gold
nanoparticles results in 57% saved energy.

At low gold loads, reduction of process window and higher required energy per section for sintering can be explained by additional energy consumption for the melting of the added gold nanoparticles. At higher loadings, energy transfer to ZnO overcompensates, so that the sintering process window increases (and minimum required energy decreases) compared to pure ZnO. Surface-adsorbed Au-NP melt down and re-solidify subsequently, together with microparticles. This is verified by TEM-pictures of FIB-sectioned microparticles in Figure 61, showing zinc oxide microparticles with 5 wt% adsorbed Au-NP before and after sintering.

Before laser sintering gold nanoparticles are homogenously distributed on the zinc oxide microparticles forming a rough surface. After laser treatment the gold nanoparticles are still visible in the TEM picture at the cross section, but seem to be embedded under a smooth and homogenous ZnO microparticle surface. Thus, a hybrid gold-semiconductor structure has been achieved.

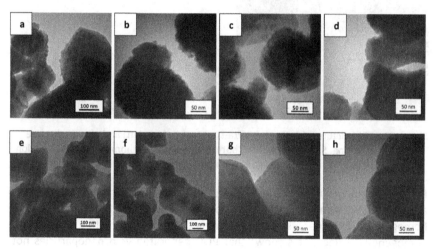

Figure 61: TEM pictures taken from the ZnO microparticles with 5 wt% Au-NP loading prepared by the FIB method (see Supplementary Information Fig. S 8-48). Particles were cut by a focused ion beam and cross-sections subsequently analyzed in a TEM. Before laser sintering surface adsorbed gold nanoparticles form a rough surface (a-d). TEM pictures taken after laser sintering of the supported micro and sub-micro particles reveal a sintering of the particles and a less rough surface after laser treatment (e-h).

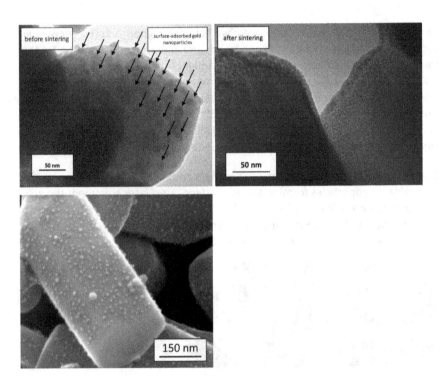

Figure 62: TEM pictures showing the ultra-structure before (left) and after (right) laser sintering. SEM picture at the bottom demonstrates homogenous distribution of laser generated gold nanoparticles supported on zinc oxide microparticles.

Beside the enhancement of laser sintering efficiency, laser irradiation of surface-adsorbed gold nanoparticles might be a method for the fabrication of homogenous micro/nano semiconductor/metal compounds with a smooth and homogenous surface. This shows that PLAL-generated gold nanoparticles not only adsorb easily on metal oxides surfaces but also that the resulting nanostructured surface adsorbate can further be influenced and post-treated by laser irradiation.

Crespo-Monteiro et al. showed that mesoporous TiO_2 films can be sintered at intensities of 97 kW/cm² with 244 nm cw laser light [Crespo-Monteiro2012]. The ablation regime was determined to start at 242 kW/cm². Sintering of TiO_2 at 488 nm did not show any material response up to 3 MW/cm² but sintering threshold could be reduced to 0.4 kW/cm², if plasmonic silver nanoparticles were added to the TiO_2. Also non-plasmonic silver oxide enabled sintering at 488 nm, what indicates that off-resonant near-field enhancement by scattering to happen in this case as well. We examined the influence of gold to IR laser sintering whereby low amounts of Au-NP reduce laser process window. 0.1 wt% of gold adsorbed on the semiconductor support results in an enhancement of process window area and decreasing thresholds for material response.

From Figure 60 the ablation threshold can be determined to be around 100 J/m for loaded and unloaded zinc oxide for spot sizes of 150 μm. If the spot diameter is decreased to 31 μm ablation threshold decreases (Supporting Information Fig. 8-49). This leads to the assumption that for high laser intensities energy transfer from gold nanoparticles can even evaporate the molten states.

Due to the observations we propose an off-resonant near-field enhancement to cause effective material melt down of ZnO microparticle surface. In case of gold nanoparticle loadings below 0.1 wt% heat transfer to the carrier material is not sufficient to melt microparticles surface. For higher loadings, Au-NP transfer enough energy to the zinc oxide. This energy is converted into heat resulting in surface melting and re-solidification leaving the sintered micro/nano hybrid structure. Figure 63 illustrates the assumed mechanism for efficiency increase of laser sintering, where near-field enhancement plays the key role.

However, it cannot be excluded that direct excitation and heating of the gold nanoparticles contributes to the heat transfer to ZnO as well. It is known that aggregation of plasmon resonant particles causes plasmon coupling leading to bathochromic shift. We expect this effect not to be dominant, since the visible absorption (color) of the gold-ZnO-material even at high loading rates does not indicate a strong SPR shift. Hence only weak absorption is expected at the applied laser wavelength. In addition it is known that particle aggregation increases their extinction coefficient mainly by (off-resonant) scattering.

Conclusions

Adsorption of gold nanoparticles, synthesized by laser ablation in liquids, can enhance laser sintering of semiconductors. If an adequate amount of gold nanoparticles is brought to the surface of the target material, significantly less (57%) energy input is required to cause a material reaction. This is of particular interest for additive manufacturing whereby lower laser power for sintering is necessary or productivity may be increased. Further, laser irradiation of the adsorbate fabricates micro/nano ZnO/Au compounds with dispersed ultra-structure. The presented experimental data present a novel approach in additive manufacturing giving access to enhanced laser sintering and at the same time surface modification of semiconductors creating metal hybrid materials. Such materials may be of interest for energy application, where light hast to be converted into current, or current into light, by combining the semiconductor bandgap with the electron capacity of the metal.

ACKNOWLEDGEMENTS

The authors thank Jurij Jakobi for TEM measurements.

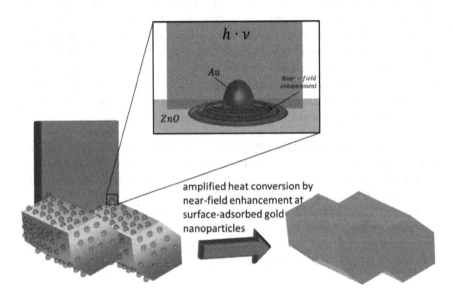

Figure 63: Schematic illustration of the supposed mechanism and material response during off-resonant near-field enhancement of infrared laser sintering. Amplification of IR laser in vicinity of the gold particle leads to energy and heat transfer to infrared transparent zinc oxide, and subsequent melting of gold nanoparticles and zinc oxide, followed by re solidification forming smooth gold-semiconductor hybrid surface.

4.3 Key figures for laser irradiation of particles in the passage reactor

This section presents a calculation of key figures based upon experimental results from PLFL in the passage reactor. These key figures might enable future studies to address mechanistic observations more precisely and may support up-scaling objectives. Here, laser parameters, particle properties and experimental design are considered to provide a comprehensive picture aiming to extract parameters that play a key role for particle processing by PLFL or PLML.

<u>Laser fluence and intensity</u>

In chapter 4.1.1, the effect of laser fluence on ps PLFL was studied. It was shown that different mechanisms occur depending on the energy density. These can be distinguished due to the experimental design of the passage reactor, as the fluence deviation within the liquid jet is small compared to reactors and experimental designs described in literature (see chapter 3.2 and Figure 13). The fluence is given by pulse energy divided by the spot area on liquid jets surface:

$$F = \frac{E_P}{A_f}$$

With: F: fluence, E_P: laser pulse energy, A_f: irradiated area on liquid jet surface (front)

Next to the laser fluence the laser intensity is a parameter used to describe the power density of a focused beam (see Figure 4). The intensity is given by the laser pulse power divided by the spot area. Dividing the pulse energy by the pulse duration gives the laser pulse power. Thus, the intensity can be determined by the following equation:

$$I = \frac{F}{\tau_P} = \frac{E_P}{\tau_P \cdot A_f}$$

With: I: intensity, τ_P: pulse duration

The fluence is colligated with the intensity by the pulse duration. However, fluence thresholds can depend on the pulse length due to different mechanisms responsible for material ablation and disruption [Chichkov1996], due to of the relaxation times of excited electrons that fall into the regime of ultra-short-pulsed lasers [Lin2003] [Zhukov2012]. Therefore, the intensity threshold does not result in a threshold valid for all pulse lengths and thus both values need to be considered, namely the intensity and fluence.

The maximum specific energy input increases with increasing laser fluence, although fragmentation efficiency is strongly limited by the optical breakdown fluence threshold of the liquid. To increase PLPPL efficiency when additional laser power is available, an increase of irradiated area can be considered as a sufficient up-scaling parameter.

Number of pulses per volume

Pulsed laser sources typically emit 10^1-10^6 pulses per second depending on the laser source and its power. This can usually be varied in a certain range. If short pulsed or ultra-short pulsed lasers irradiate a liquid flow (or when the laser beam is moved by scanning optics), laser pulses can either overlap to the previous laser pulse or leave un-irradiated areas. For laser irradiation of particle suspensions in a liquid flow, this results in a number of laser pulses that irradiate the particles while they are passing the beam line in the liquid flow. This number of pulses per volume unit (or number of pulses per particle) is given by the liquid jets' flow velocity divided by the laser spot height (equal to the spot diameter when spherical lenses are used) and the laser repetition rate:

$$N_P = \frac{h \cdot R_r}{v_l}$$

With N_P: number of pulses per volume, v_l: velocity of liquid, h: height of laser spot on liquid jets surface (note that h is equal to d_f for spherical lenses), R_r: repetition rate of pulsed laser source

This value shows whether laser pulses are overlapping (N_P>1), line up precisely (N_P=1) or are geometrically separated (N_P<1). The number of pulses per volume will be an important value if the number of particles in the irradiated area per pulse is set to 1. For values of 1, each particle will be irradiated by a single pulse only within one passage. Accordingly, further refined studies on mechanisms such particle size and morphology development are possible. A deviation of N_P

from surface of the jet to its back occurs. Exemplary calculations are also provided in chapter 4.3.1 regarding this deviation. This will result in a lower N_P value at the front to back when the spot height is higher at the surface than the back. Note that here the particle movement is regarded as static compared to the liquid flow, due to the Reynolds number of 450 (for a flow velocity of 0.35 m/s in the 1.3 mm capillary) in the liquid jet.

Relative jet illumination

To reach high energy densities and the necessary intensity, laser beams usually have to be focused for PLFL. For example, the most effective PLFL of ZnO with ps pulses at 532 nm is achieved at around 30 mJ/cm² (chapter 4.1.1). Hence, pulse energies below 150 mJ for picosecond lasers (assuming a comparably large raw beam diameter of 5 mm) requires focusing. However, this focusing will result in a smaller irradiated area compared to the unfocused beam. Here, a parameter describing the effectivity of laser excitation for particles passing the liquid jet is presented. This value sets the irradiated volume in relation to the bypassing un-irradiated volume. The combination of this parameter with the number of pulses per particle (or volume unit) provides a three-dimensional picture of particle treatment.

For up-scaling purposes, the relative jet illumination should be considered as a value that should be as close to 1 as possible. This means that no particle can pass the laser beam line un-irradiated when the number of pulses per volume is 1 or higher. In the following, it is described how this ratio can be determined by geometrical parameters of the experimental design. All that has to be known is the focal length of the optic used, the raw beam diameter, the distance of liquid jet surface to the optical focusing plane and the liquid jets' diameter. The following illustration and equations show how this can be adapted (d_f and d_B can easily be determined by linear optics).

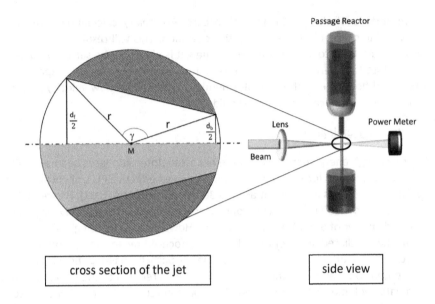

Figure 64: Schematic illustration of liquid jets cross section orthogonal to flow direction, irradiated area (50% marked green), unirradiated area (marked blue) and geometrical values used for calculation of illumination ratio (left) and the experimental set-up where this area is located (right)

Here d_f (diameter front) is the diameter of the laser spot at liquid jet front surface parallel to flow direction, d_b (diameter back) diameter of laser spot at liquid jet back surface parallel to flow direction, r radius of liquid jet, M jet center (see Figure 64). The ratio of illumination (r_{ill}) by the laser beam can be considered as the difference of the irradiated area divided by the complete cross section (A_{com}):

$$r_{ill} = \frac{A_{ill}}{A_{com}}$$

with the area of the jet cross section:

$$A_{com} = \pi \cdot r^2$$

and the area of illuminated area:

$$A_{ill} = A_{com} - A_{un}$$

and the unilluminated area (segment):

$$\frac{A_{un}}{2} = \frac{r^2}{2} \cdot (\gamma - \sin \gamma)$$

follows the term for ratio of the jet illumination:

$$r_{ill} = 1 - \frac{\gamma - \sin \gamma}{\pi}$$

Due to geometrical conditions, γ can be regarded as:

$$\gamma = 180° - \arcsin \left(\frac{d_f}{2r}\right) - \arcsin \left(\frac{d_b}{2r}\right)$$

Note that here no additional focusing due to diffraction of liquids' curved surface is considered. Nevertheless, this ratio can help to reduce the un-irradiated segments to a minimum by increasing the spot area on liquid jets' surface. When approaching the focal plane nonlinear effects should be considered, as the minimum spot diameter is given by lasers wavelength.

Furthermore, a ratio between d_f and d_b can be regarded when the focal length of used lenses is considered, which allows expressing γ simplified. Note that liquid jet radius is a fixed value given from the experimental design. The experimental studies showed that capillary diameter below 1.3 mm did not provide a continuous flow but rather formed droplets, due to the Rayleigh-Taylor instability and the Plateau-Rayleigh instability [Rayleigh1892]. Therefore this minimal capillary diameter was chosen. A larger capillary diameter would lead to significant higher flow rates and a lower ratio of illumination.

<u>Laser spot diameter at the liquid jet</u>

The spot diameter can be simply determined by using the intercept theorem. This holds in good approximation until the position of the liquid jet is within the Rayleigh length of the focus. For calculation of the laser spot at liquid jets' surface, the following equation can be used:

$$d_f = \frac{D}{f} \cdot P_{j,f}$$

With d_f: *spot diameter at liquid jets' front, D: diameter of laser beam before focusing, f: focal length of the lens, $P_{j,f}$: position of liquid jets' front as the distance to the focal plane*

Analogously, the spot at liquid jets' back can be determined by:

$$d_b = \frac{D}{f} \cdot P_{j,b}$$

With d_b: *spot diameter at liquid jets' back, $P_{j,b}$: position of liquid jets' back as distance to the focal plane*

Number of particles per pulse in irradiated volume

To determine the number of particles that are statistically in the irradiated volume, we assume a circular beam spot on liquid jets' front and back. First, the volume has to be determined and subsequently the number of particles can be calculated, when particle mass concentration and the hydrodynamic diameter of particles are known. For this determination, it is assumed that the particle size distribution comprises monomodal and ideal spheres.

The complete irradiated volume can be regarded from three different geometrical figures, namely the cone segment and the two sphere segments on the front and back of the cylindrical liquid filament. Taking the notation from the proportional jet illumination gives for the volume of cone segment (V_c):

$$V_c = \frac{2 \cdot r \cdot \pi}{3} \cdot \left(\left(\frac{d_f}{2} \right)^2 + \frac{d_f}{2} \cdot \frac{d_b}{2} + \left(\frac{d_b}{2} \right)^2 \right)$$

Here, a small volume will be neglected due to the reason that by penetration of a cone through a cylinder a section curve is formed at the front and back, whose shape depends on the ratio of the radii from the geometrical cylinder and cone to another (see Figure 8-50). The volume surrounded by this section curve will provide the accurate irradiated volume. A scheme of the conical laser beam impregnating the cylindrical liquid jet is shown in Figure 8-50. Therefore, assuming a cone segment with planar surfaces will slightly overestimate the illuminated volume. The maximum possible error hereby is below 22% (if the beam spots at liquid jets' front and back are larger than liquid jets' diameter) and becomes neglectable if the spot diameter is in the range of liquid jets' radius or smaller.

In the following, a correlation between particle mass concentration, particle diameter (assuming monomodal spheres) and the illuminated volume is drawn. The particle mass concentration can be described as follows:

$$wt\% = \frac{m_{p,ges}}{m_{ges}} \cdot 100$$

The mass (in irradiated volume) of particles $m_{p,ges}$ is:

$$m_{p,ges} = n_p \cdot V_p \cdot \rho_p$$

With n_p: *number of particles*, V_p: *volume of a single particle*, ρ_p: *density of particles (bulk value)*

The volume of a single (spherical) particle can be expressed by:

$$V_p = \frac{1}{6}\pi \cdot d_p{}^3$$

With d_p: *diameter of the particle*

The mass of irradiated volume m_{ges} is:

$$m_{ges} = \left(V_{com} - V_p \cdot n_p\right) \cdot \rho_l + n_p \cdot V_p \cdot \rho_p$$

With V_{com}: *volume of the complete irradiated volume*, ρ_l: *densitry of liquid*

Using the equation for m_{ges} and $m_{p,ges}$ to plug into the equation for particle mass concentration and solving the equation for the number of particles (n_p) gives:

$$n_p = \frac{V_c \cdot \rho_l}{\dfrac{V_p \cdot \rho_p \cdot 100}{wt\%} + V_p \cdot \rho_l - V_p \cdot \rho_p}$$

With this equation, the number of educt particles statistically irradiated per laser pulse can be determined by knowledge of particle mass concentration in liquid, particle size (assuming monomodal spheres, this value can be derived from particle size distribution), laser spot diameters at front and back of the liquid jet, as well as the liquid jet diameter.

The key figures N_P (number of pulses per volume), r_{ill} (ratio of illumination) and n_p (statistical number of particles in irradiated volume) can be used for further up-scaling purposes and precise design of experimental conditions, e.g. addressing single-pulse-per-particle passages.

4.3.1 Exemplary calculations for PLFL of ZnO

To demonstrate how these calculations can be applied, the optimized conditions and parameters experimentally derived in this thesis for PLFL of ZnO are considered. For this purpose, the picosecond Nd:YAG laser was used at the second-harmonic wavelength (532 nm) with a repetition rate of 100 kHz. The raw beam diameter was determined experimentally to be 3.5 mm. Optimized fragmentation efficiency was found to be close to the optical breakdown at around 0.1 J/cm² equivalent to 100 GW/cm² for 10 ps. By using a 100 mm plano convex focal lens, the optical breakdown occurs for a distance of around 91 mm from the lens to the surface of the liquid jet. Setting the liquid jet surface to a distance of 89 mm from the lens gives optimized fragmentation conditions, resulting in a calculated spot diameter of 0.388 mm at the front and 0.353 mm at the back.

The velocity of the liquid jet was determined by measuring the mass of water outflow of the reactor each 10 seconds. Dividing this volume flow rate by the area of the jet's cross section gives the velocity, as illustrated in Figure 8-51. For the experiments, a volume of 50 ml was used and thus the starting velocity of around 0.35 m/s can be estimated. Average velocity for 50 ml passaging (0.35 m/s – 0.325 m/s) is 0.338 m/s with a deviation of 4%. This deviation increases linearly with the volume used for a passage (up to 26% for 350 ml). This can be technically compensated by using the principle of Mariotte's bottle [McCarthy1934].

The following table summarizes all parameters necessary for the calculations.

Table 4: Parameters for optimized PLFL of ZnO necessary for calculation of the key figures

Parameter	variable	Value
Liquid jet parameter		
particle concentration in liquid	$wt\%$	0.1 wt%
liquid jet velocity	v_l	0.3375 m/s (±0.0125 m/s)
radius of the liquid jet	r	0.65 mm
position of liquid jets' front as distance to the focal plane	$P_{j,f}$	11 mm
position of liquid jets' back as distance to the focal plane	$P_{j,b}$	9.7 mm
Laser parameter		
repetition rate of the laser	R_r	100 kHz
beam diameter before focusing	D	3.5 mm
focus length of the lens	f	100 mm
pulse energy	E_P	75 μJ
Colloid parameter		
hydrodynamic educt particle diameter	d_p	450 nm
density of the liquid (water)	ρ_l	1 g/cm³
Density of the particles (ZnO)	ρ_p	5.61 g/cm³

For the laser spot diameter at the liquid jet

$$d_f = \frac{D}{f} \cdot P_{j,f} = \frac{3.5\ mm}{100\ mm} \cdot 11\ mm = 0.385\ mm$$

$$d_b = \frac{D}{f} \cdot P_{j,b} = \frac{3.5\ mm}{100\ mm} \cdot 9.7\ mm = 0.3395\ mm$$

For the number of pulses per volume:

$$N_P = \frac{h \cdot R_r}{v_l} = \frac{0.000385 \ m \cdot 100{,}000 \ s^{-1}}{0.3375 \frac{m}{s}} \approx 114.07$$

Therefore, 114 laser pulses irradiate the particle suspension volume passing the liquid jet.

The ratio of jet illumination can be described as:

$$\gamma = 180° - \arcsin\left(\frac{d_f}{2r}\right) - \arcsin\left(\frac{d_b}{2r}\right)$$
$$= 180° - \arcsin\left(\frac{0.385 \ mm}{1.3 \ mm}\right) - \arcsin\left(\frac{0.3395 \ mm}{1.3 \ mm}\right)$$
$$\gamma = 180° - 17.23° - 15.14° = 147.63° \triangleq 0.82\pi \ rad$$

Please note that the equation for the ratio of illumination has to be solved in radiant:

$$r_{ill} = 1 - \frac{\gamma - \sin\gamma}{\pi} = 1 - \frac{2.58 - 0.535}{\pi} = 0.35$$

Thus, 35% of the liquid jet cross section is illuminated.

The number of particles in the laser-excited volume can be determined by the following equations:

The volume for the cone segment is given by:

$$V_c = \frac{2 \cdot r \cdot \pi}{3} \cdot \left(\left(\frac{d_f}{2}\right)^2 + \frac{d_f}{2} \cdot \frac{d_b}{2} + \left(\frac{d_b}{2}\right)^2 \right)$$

$$V_c =$$

$$\frac{1.3 \, mm \cdot \pi}{3} \cdot \left(\left(\frac{0.385 \, mm}{2}\right)^2 + \frac{0.385 \, mm}{2} \cdot \frac{0.3395 \, mm}{2} + \left(\frac{0.3395 \, mm}{2}\right)^2 \right)$$

$$\approx 0.134 \, mm^3$$

The volume of a single particle is:

$$V_p = \frac{1}{6}\pi \cdot d_p{}^3 = \frac{1}{6}\pi \cdot (450 \, nm)^3 = 0.0477129 \, \mu m^3$$

Thus, the number of (educt particles) in the irradiated area is:

$$n_p = \frac{V_c \cdot \rho_l}{\frac{V_p \cdot \rho_p \cdot 100}{wt\%} + V_p \cdot \rho_l - V_p \cdot \rho_p}$$

$$n_p =$$

$$\frac{0.134 \, mm^3 \cdot 1\frac{g}{cm^3}}{\frac{(0.0477129 \cdot 10^{-9}) \, mm^3 \cdot 5.61\frac{g}{cm^3} \cdot 100}{0.1 \, wt\%} + (0.0477129 \cdot 10^{-9}) \, mm^3 \cdot 1\frac{g}{cm} - (0.0477129 \cdot 10^{-9}) \, mm^3 \cdot 5.61\frac{g}{cm^3}}$$

$$= 501,029$$

Therefore, around half a million particles are illuminated in the irradiated area with about 114 laser pulses.

Table 5: Parameters of a single passage determined for optimized PLFL conditions of ZnO

number of pulses per particle	114
ratio of jet illumination	0.35
Volume of illuminated liquid	0.134 mm³
number of particles in irradiated volume	501,029

To summarize, it was shown that the key figures can be used to calculate and estimate process conditions, using the parameters reported in chapter 4.1 for PLFL of ZnO.

4.3.2 Optimization of PLFL using key figures

Optimized conditions can be derived by correlation of the key figures with the suspension and laser parameters, which holds particular interest for up-scaling and to reduce the required number of passages. For picosecond pulses at 532 nm wavelength, the threshold fluence of optical breakdown of the liquid was found to be around 0.1 J/cm². To achieve a higher fragmentation efficiency, either the particle concentration can be reduced (as described in chapter 4.2) or the illuminated volume at a constant fluence can be increased, although for this higher laser power (pulse energy) needs to be available. The following diagram illustrates the relative jet illumination in dependence of the applied laser fluence for different laser power. For this calculation, the 1.3 mm liquid jet diameter, 10 picosecond pulse duration with 532 nm wavelength, 100 kHz repetition rate and a focusing lens of 100 mm were assumed.

Figure 65 shows three different diagrams. In Figure 65 a) the laser fragmentation efficiency is plotted for different fluences on the liquid jets surface. The effective PLFL regime is marked similar to Figure 65 b) and c). In Figure 65 b) and c) the correlation of the relative jet illumination and the laser fluence on the liquid jets surface is plotted. This correlation plotted in double logarithmic scale for a wide range of fluences on the left (b)) and a magnification of the PLFL regime on the right (c)).

The blue curve in Figure 65 a) correlates to the blue curve in Figure 65 b) and c). Therefore, only around 35% to 20% of the liquid are illuminated during one passage.

For fixed focusing and liquid jet geometries, a direct correlation between the irradiated volume and the ratio of illumination can be plotted. This is shown in Figure 8-53 for the experimental conditions used. Accordingly, the laser fluence can also be plotted in correlation with the ratio of illumination, as shown in Figure 65 (analogous to Figure 8-52). This illustration is more intuitive as a ratio of 1 is a 100 % illumination of the liquid jet in diameter. The optimized fragmentation conditions for 0.1 wt% ZnO were found at a laser fluence of 0.03 J/cm². As the fluence scales linearly with the pulse energy, a 100 % illumination with a fluence of 0.03 J/cm² can be reached by a pulse energy of 1.5 mJ (150 watts at 100 kHz repetition rate). Therefore, 20 times higher laser pulse energy than 75 µJ used in the present study is desired to reach the fluence of 0.03 J/cm² at an irradiated volume of around 7 mm³ equal to the total excitation of the jet volume. Note that with 750 µJ 10 times higher than the pulse energy used the PLFL regime can also be addressed with a complete illumination of the liquid jet. However, PLFL becomes more efficient for fluences up to 0.03 J/cm² (see Figure 16). Higher energy densities result in a decrease of PLFL efficiency until a complete optical breakdown of the liquid jet occurs (see Figure 16).

If laser pulse energy higher than 1.5 mJ is accessible, other geometries of focusing lenses can be considered to further increase the irradiated volume, such as cylindrical lenses. Hence, the bypassing flow of untreated educt particles causing educt-product mixing could be avoided. It is assumed that such educt-product mixing reduces the PLFL efficiency, e.g. if product nanoparticles adsorb on the educt particle surface, thus increasing the educt´s hydrodynamic diameter and light absorption cross section during subsequent passages. Note that no surfactants are required to benefit from one of the key advantages of PLFL, namely its purity. Hence, using this key figure to adjust a relative jet illumination of 100% may contribute to a further increase in PLFL efficiency. Note that laser sources providing this pulse energy are already available.

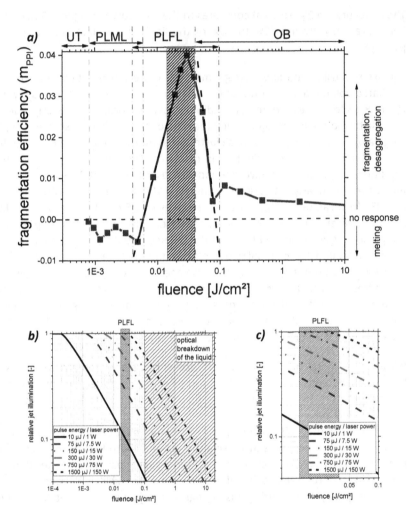

Figure 65: a) Laser fragmentation efficiency (see Figure 16) for the experimentally used picosecond laser parameters (75 µJ, 100 kHz, 532 nm and 10 ps) b) Relative jet illumination plotted versus the fluence for different laser power assuming a liquid jet diameter of 1.3 mm, a laser raw beam diameter of 3.5mm and a focal length of 100 mm, 10 ps pulse length, 532 nm wavelength, 100 kHz repetition rate and 0.1 wt% particle mass concentration and c) magnification of the PLFL regime

It can be stated that mainly focusing geometry, laser parameters and suspension properties determine the process efficiency. Here, it is shown in detail how the process can be further optimized, exemplified by the optimized experimental conditions for zinc oxide. To transfer this optimization procedure to other materials, fragmentation thresholds should be determined. Subsequently, the optimization can be easily transferred to other particle/liquid systems. The determination of these parameters and their conjunction was possible due to the development of the passage reactor and the experimental studies of PLFL using this design.

An optimized reconciliation of all parameters such as laser parameters, colloid parameters and liquid jet parameters might result in a single passage from educt to product, although this surely depends on desired properties such as product particle size or product defect properties.

5 Summary and conclusion

Pulsed laser processing of particles in liquid is a technique giving access to a variety of highly interesting nano-particulate and sub-micrometer spherical materials with unique properties, which would be difficult or impossible to fabricate using conventional synthesis methods. The demand contributing in an additional mechanistic understanding of PLPPL derives from literature reports stating different mechanistic assumptions and demonstrating the possibility of chemical conversion. But at present, PLPPL is conducted by continuous irradiation of particle suspensions in a vessel resulting in a significant fluence gradient during beam propagation through the liquid. The experimental studies show that a precise experimental design is required for energy-efficient and meticulous laser fragmentation or melting conditions. For this, based on the idea of a free liquid jet introduced by Wagener et al. [Wagener2010a], [Wagener2011], the experimental design was improved and changed to a passage reactor enabling sequential laser irradiation with minimal laser fluence deviation. This is of importance, as the process windows for PLPPL are limited at low and high laser fluences for laser melting and fragmentation, respectively.

For picosecond laser pulses and 532 nm laser wavelength particle fragmentation in liquids was extensively studied. The sequential laser fragmentation of particles in the passage reactor enables energy balancing and correlation of the specific energy input to particle properties for the first time. The investigations on zinc oxide particles formed by laser fragmentation under optimized fragmentation conditions show that the produced particles are highly defect-rich. Investigating the bandgap energy of the defect-rich ZnO particles revealed that their band gap can be adjusted linearly by the laser energy dose put into the educt microparticles. Such defect control could be interesting for future catalysis research. Furthermore, laser fragmentation allows even higher extend of chemical modification of the generated particles. This has been demonstrated by reduction of copper-(II)-oxide and copper nitride edcut particles to elemental copper nanoparticles during fragmentation in ethyl acetate, ethanol or isopropanol. Oxidation of the copper nanoparticles can be observed using water as solvent. Laser fragmentation in presence of an oxidative reagent such as hydrogen peroxide can fabricate ultra-small and ligand-free gold nanoparticles around 2.5 nm in diameter. Consequently, ultra-small ligand-free gold nanoparticles in the quantum size regime, have been synthesized for the first time. An activation of particles for subsequent oxidation is also possible as shown for aluminum microparticles that are treated by laser fragmentation in water, whereby subsequent hydrogen release potentially from water splitting can be observed. Hence, combination of liquid medium and particle material determine process chemistry, showing that both reductive (defects) and oxidative PLPPL is possible.

Longer laser pulse lengths and shorter wavelengths, such as nanoseconds and 355 nm, can cause distinctive particle melting and formation of sub-micrometer spheres. Utilizing the experimental design of the passage reactor shows that this setup can be used for melting condition and process window determination as well. This was studied for zinc oxide particles and the findings of melting fluence thresholds are in agreement with the literature. Only the generation of multi material sub-micrometer spheres remains barely reported in literature and all related reports are precursor chemistry-based. In this work it is shown how metal-semiconductor-composite sub-micrometer spheres can be fabricated by pulsed laser melting in pure water for the first time. Here, plasmonic gold nanoparticles are integrated into their zinc oxide support, forming crystalline metallic particles integrated in the semiconductor sub-micrometer spheres.

For pulsed laser melting in liquids the presence of gold nanoparticles on microparticles' surface seems to have no strong impact on the process efficiency. On the contrary, in the case of laser additive manufacturing of such pasmonic/semiconductor hybrid materials in the dry state (powder bed), a strong influences on the process can be observed. The experimental studies demonstrate that this can result in widening of the laser sintering process window if gold nanoparticle are present during the laser structuring process. The plasmoic response of gold nanoparticles was found to be responsible for the observed effects within different nanoparticle loadings used. Besides extending the laser structuring process window, an ultra-structure of gold nanoparticles dispersed in a solid ZnO matrix could be generated out of the particulate powder bed.

For PLPPL in the passage reactor key figures and formulas balancing the specific process parameters are derived from the experimental results, which will enable up-scaling of the laser fragmentation process in a free liquid jet. This geometric modelling allows a precise parameter definition. In general, we can conclude that for most effective laser fragmentation, the laser fluence should be as high as possible but just below the optical breakdown of the liquid. The number of pulses required to reach the desired particle properties, such as particle size, composition or defects, can be determined by this unique passage irradiation method and used to scale the process. It has been exemplarily calculated for ps laser fragmentation how laser parameters affect the ratio of illumination. These geometrical conditions can be used to set the jet illumination to 1, thus avoiding particle passing in the liquid jet without laser irradiation. The number of particles in the irradiated volume depends on the particle concentration, educt particle size and can be adjusted easily by variation of the particle mass concentration.

In conclusion, it can be stated that the developed passage reactor design is an appropriate setup for experimental studies to understand PLPPL. Next to the experimental studies, this technique is easily scalable and key figures can be applied when using other laser systems to achieve the same results. The results are in agreement with the existing literature and contribute significantly to a new process understanding and modeling. The valuable model parameter that have been derived from PLPPL in the passage reactor allows to further improve this technique and will be significant going forward as defined setting of nanoparticle properties is possible.

6 Outlook

The dissertation addresses fundamental aspects of PLPPL using a novel experimental design. This improved experimental design enables determination of particle properties such as the minimal obtainable particle size caused by pulsed laser fragmentation in liquids and the yield during PLPPL. It might be projected that besides the laser fluence and optical properties the educt particle concentration has an influence on the mechanisms, due to attenuation of laser light in the liquid occurring in case of high concentrations. Detailed studies correlating the dominant mechanisms such as fragmentation and/or vaporization to optical absorption properties of the particles and studying fluence regimes will help in additional and comprehensive understanding. The implementation of the passage reactor for single particle irradiated with a single laser pulse, might be appropriate to further gain insight on mechanisms occurring during pulsed laser fragmentation in liquids. Detailed characterization of the particle size and its composition after laser irradiation with a single pulse could further contribute in a mechanistic understanding.

Although PLFL-chemistry has been shown it remains open which parameters set the final chemical composition of particles by pulsed laser melting in liquids. For this, copper nitride and copper oxide are good candidates because pulsed laser fragmentation of these particle materials showed formation of copper nanoparticles. Thus, when exposed to precise laser fluences and wavelengths known to cause sufficient melting additional mechanistic conclusions could be drawn. Exposing Cu_3N and CuO to laser fluences between 30 mJ/cm² and 150 mJ/cm² and subsequent analysis of sub-micrometer sphere composition can give scientific evidence if their composition changes or if higher fluences are required for the chemical conversion.

Up-scaling of the process by scanning the laser beam along the liquid jet following the flow direction might be one option to reduce the required numbers of irradiation cycles (passages). Setting the scan speed of the laser at least one order of magnitude faster that the liquid jet flow velocity could further increase the process efficiency. Increasing the number of pulses by higher laser repetition rates might also reduce the required number of passages. But as nanobubble formation with lifetimes of around 1 ns and 50 nm diameter (for 100 femtosecond pulses and 160 mJ/cm^2) are reported for gold nanoparticles [Kotaidis2005] there might also be a limitation in the repetition rates due to shielding effects, if the bubble lifetime increases significantly. 80 ns bubble lifetimes (for 500 picosecond pulses and 220 mJ/cm^2) were reported from Hashimoto et al. [Hashimoto2012]. This effect might be even more distinctive for PLML due to larger particles and longer pulse duration of ns compared to ps. For optimal illumination the relative illumination should be set to 1, meaning that the complete liquid filament in diameter is irradiated to avoid particles passing the laser beam unirradiated.

Optimization of the PLPPL process could generate the desired particle properties in 1 passage. The parameters (extracted from experiments) required for an optimized processing are reported in this work.

7 Literature

[Agarwala1995]   M. Agarwala, D. Bourell, J. Beaman, H. Marcus, J. Barlow, *Direct selective laser sintering of metals*, Rapid Prototyping J. 1 (**1995**) 26-36

[Ahmad2011]   M. Ahmad, S. Yingying, A. Nisar, H. Sun, W. Shen, M. Weie, J. Zhu, *Synthesis of hierarchical flower-like ZnO nanostructures and their functionalization by Au nanoparticles for improved photocatalytic and high performance Li-ion battery anodes*, J. Mater. Chem. 21 (**2011**) 7723-7729

[Akita2000]   S. Akita, H. Nishijima, Y. Nakayama, *Influence of stiffness of carbon-nanotube probes in atomic force microscopy*, J. Phys. D: Appl. Phys. 33 (**2000**) 2673-2677

[Aitken1881]   J. Aitken, *On the colour of the Mediterranean and other waters*, Proceedings of the Royal Society of Edinburgh (**1881–1882**) 11:472–483

[Amendola2007]   V. Amendola, M. Meneghetti, *Controlled size manipulation of free gold nanoparticles by laser irradiation and their facile bioconjugation*, J. Mater. Chem. 17 (**2007**) 4705–4710

[Amendola2008]   V. Amendola, *Synthesis of gold and silver nanoparticles for photonic applications*, PhD Thesis, University of Padova, Italy, **2008**

[Amendola2009]   V. Amendola, M. Meneghetti, *Laser ablation synthesis in solution and size manipulation of noble metal nanoparticles*, Phys. Chem. Chem. Phys. 11 (**2009**) 3805–3821

[Amendola2013]   V. Amendola, M. Meneghetti, *What controls the composition and the structure of nanomaterials generated by laser ablation in liquid solution?*, Phys. Chem. Chem. Phys. 15 (**2013**) 3027-3046

[Asahi2008]   T. Asahi, T. Sugiyana, H. Masuhara, *Laser Fabrication and Spectroscopy of Organic Nanoparticles*, Acc. Chem. Res. 41 (**2008**) 1790-1798

[Asahi2015]   T. Asahi, F. Mafune, C. Rehbock, S. Barcikowski, *Strategies to harvest the unique properties of laser-generated nanomaterials in biomedical and energy applications*, Appl. Surf. Sci 348 (**2015**) 1-3

[Avadhut2012]   Y. Avadhut, J. Weber, E. Hammarberg, C. Feldmann, J. Schmedt auf der Günne, *Structural investigation of aluminium doped ZnO nanoparticles by solid-state NMR spectroscopy*, Phys. Chem. Chem. Phys. 14 (**2012**) 11610-11625

[Awazu2008]   K. Awazu, M. Fujimaki, C. Rockstuhl, J. Tominaga, H. Murakami, Y. Ohki, N. Yoshida, T. Watanabe, *A Plasmonic Photocatalyst Consisting of Silver Nanoparticles Embedded in Titanium Dioxide*, J. Am. Chem. Soc. 130 (**2008**) 1676-1680

[Badr2007]   Y. Badr, M.A. Mahmoud, *Excimer laser photofragmentation of metallic nanoparticles*, Phys. Lett. A 370 (**2007**) 158–161

[Bai2006]   S. Bai, T. Tseng, *Effect of alumina doping on structural, electrical, and optical properties of sputtered ZnO thin films*, Thin Solid Films 515 (**2006**) 872-875

[Baersch2009]   N. Baersch, J. Jakobi, S. Weiler, S. Barcikowski, *Pure colloidal metal and ceramic nanoparticles from high-*

power picosecond laser ablation in water and acetone, Nanotechnology 20 (**2009**) 445603 (9pp)

[Balandin2013]   P. Balandin, K. Maximova, M. Gongalsky, J. Sanchez-Royo, V. Chirvony, M. Sentis, V. Timoshenko, A. Kabashin, *Femtosecond laser fragmentation from water-dispersed microcolloids: toward fast controllable growth of ultrapure Si-based nanomaterials for biological applications,* J. Mater. Chem. B 1 (**2013**) 2489-2495

[Barchanski2011]   A. Barchanski, C. Sajti, C. Sehring, S. Petersen, S. Barcikowski, *Design of Bi-functional Bioconjugated Gold Nanoparticles by Pulsed Laser Ablation with Minimized Degradation,* J. Laser Micro Nanoen. 6 (**2011**) 124–130

[Barchanski2012]   A. Barchanski, N. Hashimoto, S. Petersen, C. Sajti, S. Barcikowski, *Impact of Spacer and Strand Length on Oligonucleotide Conjugation to the Surface of Ligand-Free Laser-Generated Gold Nanoparticles,* Bioconjugate Chem. 23 (**2012**) 908–915

[Barcikowski2009]   S. Barcikowski, F. Devesa, K. Moldenhauer, *Impact and structure of literature on nanoparticle generation by laser ablation in liquids,* J. Nanopart. Res. 11 (**2009**) 1883–1893

[Barcikowski2009a]   S. Barcikowski, J. Walter, A. Hahn, J. Koch, H. Haloui, T. Herrmann, A. Gatti, *Picosecond and Femtosecond Laser Machining May Cause Health Risks Related to Nanoparticle Emission,* J. Laser Micro Nanoen. 4 (**2009**) 159-164

[Barcikowski2013]   S. Barcikowski, G. Compagnini, *Advanced nanoparticle generation and excitation by lasers in liquids,* Phys. Chem. Chem. Phys. 9 (**2013**) 3022-3026

[Behrens2015]   M. Behrens, *Coprecipitation: An excellent tool for the synthesis of supported metal catalysts – From the understanding of the well known recipes to new materials*, Catal. Today 246 (**2015**) 46-54

[Besner2006]   S. Besner, A. Kabashin, M. Meunier, *Fragmentation of colloidal nanoparticles by femtosecond laser-induced supercontinuum generation*, Appl. Phys. Lett. 89 (**2006**) 233122 (2pp)

[Besner2010]   S. Besner, M. Meunier, *Femtosecond Laser Synthesis of AuAg Nanoalloys: Photoinduced Oxidation and Ions Release*, J. Phys. Chem. C 114 (**2010**) 10403–10409

[Bonaccorso2013]   F. Bonaccorso, M. Zerbetto, A. C. Ferrari, V. Amendola, *Sorting Nanoparticles by Centrifugal Fields in Clean Media*, J. Phys. Chem. C 117 (**2013**) 13217-13229

[Boulais2012]   É. Boulais, R. Lachaine, M. Meunier, *Plasma Mediated off-Resonance Plasmonic Enhanced Ultrafast Laser-Induced Nanocavitation*, Nano Lett. 12 (**2012**) 4763-4769

[Brust1994]   M. Brust, M. Walker, D. Bethell, D. J. Schiffrin, R. Whyman, *Synthesis of Thiol-derivatised Gold Nanoparticles in a Two-phase Liquid-Liquid System*, Chem. Commun. (**1994**) 801-802

[Cai1998]   H. Cai, N. Chaudhary, J. Lee, M. Becker, J. Brock, J. Keto, *Generation of metal nanoparticles by laser ablation of microspheres*, J. Aerosol Sci. Vol. 29 (**1998**) 627–636

[Cavicchi2013]   R. Cavicchi, D. Meier, C. Presser, V. Prabhu, S. Guha, *Single Laser Pulse Effects on Suspended-Au-Nanoparticle Size Distributions and Morphology*, J. Phys. Chem. C 117 (**2013**) 10866–10875

[Chanu2014]   T. Chanu, T. Muthukumar, P. Manoharan, *Fuel mediated solution combustion synthesis of ZnO supported gold clusters and nanoparticles and their catalytic activity and in vitro cytotoxicity*, Phys. Chem. Chem. Phys. 16 (**2014**) 23686-23698

[Chen2001]   Y. Chen, C. Yeh, *A new approach for the formation of alloy nanoparticles: laser synthesis of gold–silver alloy from gold–silver colloidal mixtures*, Chem. Commun. (**2001**) 371–372

[Chichkov1996]   B. Chichkov, C. Momma, S. Nolte, F. Alvensleben, A. Tuennermann, *Femtosecond, picosecond and nanosecond laser ablation of solids*, Appl. Phys. A 63 (**1996**) 109-115

[Chithrani2007]   B. Chithrani, W. Chan, *Elucidating the Mechanism of Cellular Uptake and Removal of Protein-Coated Gold Nanoparticles of Different Sizes and Shapes*, Nano Lett. 7 (**2007**) 1542-1550

[Chubilleau2011] C. Chubilleau, B. Lenoir, S. Migot, A. Dauscher, *Laser fragmentation in liquid medium: A new way for the synthesis of PbTe nanoparticles*, J. Colloid Interface Sci. 357 (**2011**) 13–17

[Ciupina2004]   V. Ciupina, I. Carazeanu, G. Prodan, *Characterization of $ZnAl_2O_4$ Nanocrystals prepared by coprecipiration and microemulsion techniques*, J. Optoelectron. Adv. M. 6 (**2004**) 1317 – 1322

[Crespo-Monteiro2012] N. Crespo-Monteiro, N. Destouches, L. Saviot, S. Reynuad, T. Epicier, E. Gamet, L. Bios, A. Boukenter, *One-Step Microstructuring of TiO2 and Ag-TiO2 Films by Continuous Wave Laser Processing in the UV and Visible Ranges*, J. Phys. Chem. C 116 (**2012**) 26857-26864

[Daniel2004]   M.-C. Daniel, D. Astruc, *Gold Nanoparticles: Assembly, Supramolecular Chemistry, Quantum-Size-Related Properties, and Applications toward Biology, Catalysis, and Nanotechnology*, Chem. Rev. 104 (**2004**) 293-346

[Degen2000]   A. Degen, M. Kosec, *Effect of pH and impurities on the surface charge of zinc oxide in aqueous solution*, J. Eur. Ceram. Soc. 20 (**2000**) 667-673

[DeGiacomo2013]   A. De Giacomo, M. Dell'Aglio, A. Santagata, R. Gaudiuso, O. De Pascale, P. Wagener, G. C. Messina, G. Compagnini, S. Barcikowski, *Cavitation dynamics of laser ablation of bulk and wire-shaped metals in water during nanoparticles production*, Phys. Chem. Chem. Phys. 15 (**2013**) 3083-3092

[Dhas2008]   V. Dhas, S. Muduli, W. Lee, S. Han, S. Ogale, *Enhanced conversion efficiency in dye-sensitized solar cells based on ZnO bifunctional nanoflowers loaded with gold nanoparticles*, Appl. Phys. Lett. 93 (**2008**) 243108 (2pp)

[Dijken2000a]   A. van Dijken, E. Meulenkamp, D. Vanmaekrlbergh, A. Meijerink, *The luminescence of nanocrystalline ZnO particles: the mechanism of the ultraviolet and visible emission*, J. Lumin. 87-89 (**2000**) 454-456

[Dijken2000b]   A. van Dijken, E. Meilenkamp, D. Vanmaekelbergh, A. Meijerink, *Identification of the transition responsible for the visible emission in ZnO using quantum size effects*, J. Lumin. 90 (**2000**) 123-128

[Dorranian2012]   D. Dorranian, E. Solati, L. Dejam, *Photoluminescence of ZnO nanoparticles generated by laser ablation in deionized water*, Appl. Phys. A 109 (**2012**) 307-314

[Dudziak2010]   S. Dudziak, M. Gieseke, H. Haferkamp, S. Barcikowski, D. Kracht, *Functionality of laser-sintered shape memory micro-actuators*, Phys. Procedia 5 (**2010**) 607-615

[Ellmer2001]   K. Ellmer, *Resistivity of polycrystalline zinc oxide films: current status and physical limit,* J. Phys. D: Appl. Phys. 34 (**2001**) 3097–3108

[Elzey2011]   S. Elzey, J. Baltrusaitis, S. Bian, V. Grassian, *Formation of paratacamite nanomaterials via the conversion of aged and oxidized copper nanoparticles in hydrochloric acidic media*, J. Mater. Chem. 21 (**2011**) 3162

[Faraday1857]   M. Faraday, *The Bakerian Lecture: Experimental Relations of Gold (and Other Metals) to Light*, Phil. Trans. R. Soc. Lond. 147 (**1857**) 145-181

[Fojtik1993]   A. Fojtik, A. Henglein, *Laser ablation of films and suspended particles in a solvent: formation of cluster and colloid solutions*, Ber. Bunsenges. Phys. Chem. 97 (**1993**) 252–254

[Fujiwara2013]   H. Fujiwara, R. Niyuki, Y. Ishikawa, N. Koshizaki, T. Tsuji, *Low-threshold and quasi-single-mode random laser within a submicrometer-sized ZnO spherical particle film*, Appl. Phys. Lett. 102 (**2013**) 061110 (4pp)

[Furukawa1968]   G. Furukawa, W. Saba, M. Reilly, "*Critical Analysis of the Heat-Capacity Data of the Literature and Evaluation of Thermodynamic Properties of Copper, Silver, and Gold from 0 to 300 K*," U.S. Dept. of Commerce, National Bureau of Standards (U.S. Government Printing Office, Washington, D.C., **1968**)

[Garcia2005]   M. Garcia, J. Venta, P. Crespo, J. LLoips, S. Penadés, A. Fernández, A. Hernando, *Surface plasmon resonance of capped Au nanoparticles*, Phys. Rev. B 72 (**2005**) 241403

[Garcia2014]   L. García, M. Mendivil, T. Das Roy, G. Castillo, S. Shaji, *Laser sintering of magnesia with nanoparticles of iron oxide andaluminum oxide*, Appl. Surf. Sci. 336 (**2014**) 59-66

[Gedzelman2005]   S. Gedzelman, *Simulating colors of clear and partly cloudy skies*, Appl. Optics 44 (**2005**) 5723-5736

[Giammanco2010]   F. Giammanco, E. Giorgetti, P. Marsili, A. Giusti, *Experimental and Theoretical Analysis of Photofragmentation of Au Nanoparticles by Picosecond Laser Radiation*, J. Phys. Chem. C 114 (**2010**) 3354–3363

[Giljohann2010]   D. Giljohann, D. Seferos, W. Daniel, M. Massich, P. Patel, C. Mirkin, *Gold Nanoparticles for Biology and Medicine*, Angew. Chem. Int. Ed. 49 (**2010**) 3280–3294

[Giorgetti2014]   E. Giorgetti, P. Marsili, M. Muniz-Miranda, C. Gellini, F. Giammanco, *Spectroscopic evidence of positive clusters in Ag colloids obtained by laser ablation in aqueous solutions*, Appl. Phys. A 117 (**2014**) 327–331

[Gökce2015]   B. Gökce, S. Barcikowski, P. Behrens, U. Fritsching, I. Kelbassa, R. Poprawe, C. Esen, A. Ostendorf, B. Voit, *Prozessadaptierte Materialien für die Photonik*, Photonik 47 (**2015**) 24-28

[Gökce2015a]   B. Gökce, D. van't Zand, A. Menendez-Manjon, S. Barcikowski, *Ripening kinetics of laser-generated plasmonic nanoparticles indifferent solvents*, Chem. Phys. Lett. 626 (**2015**) 96–101

[Goesmann2010]   H. Goesmann, C. Feldmann, *Nanopartikuläre Funktionsmaterialien*, Angew. Chem. 122 (**2010**) 1402 – 1437

[Goldys2012]   E. Goldys, M. Sobhan, *Fluorescence of Colloidal Gold Nanoparticles is Controlled by the Surface Adsorbate*, Adv. Funct. Mater. 22 (**2012**) 1906-1913

[Gu2012]   D. Gu, W. Meiners, K. Wissenbach, R. Poprawe, *Laser additive manufacturing of metallic components: materials, processes and mechanisms*, Int. Mater. Rev. 57 (**2012**) 133-164

[Gupta1990]   T. Gupta, *Application of Zinc Oxide Varistors*, J. Am. Ceram. Soc. 73 (**1990**) 1817-1840

[Häkkinen2012]   H. Häkkinen, *The gold–sulfur interface at the nanoscale*, Nat. Chem. 4 (**2012**) 443-455

[Hahn2010]   A. Hahn, T. Stöver, G. Paasche, M. Löbler, K. Sternberg, H. Rohm, S. Barcikowski, *Therapeutic Window for Bioactive Nanocomposites Fabricated by Laser*

Ablation in Polymer-Doped Organic Liquids, Adv. Eng. Mater. 12 (**2010**) 156-162

[Haiss2007] W. Haiss, N. Thanh, J. Aveyard, D. Fernig, *Determination of Size and Concentration of Gold Nanoparticles from UV-Vis Spectra*, Anal. Chem. 79 (**2007**) 4215-4221

[Haruta1987] M. Haruta, T. Kobayashi, H. Sano, N. Yamada, *Novel Gold Catalysts for the Oxidation of Carbon Monoxide at a Temperature far Below 0°C*, Chem. Lett. (**1987**) 405-408

[Haruta1997] M. Haruta, *Size- and support-dependency in the catalysis of gold*, Catal. Today 36 (**1997**) 153-166

[Hashmi2006] S. Hashmi, G. Hutchings, *Gold Catalysis*, Angew. Chem. Int. Ed. 45 (**2006**) 7896-7936

[Hashimoto2012] S. Hashimoto, D. Werner, T. Uwarda, *Studies on the interaction of pulsed lasers with plasmonic gold nanoparticles toward light manipulation, heat management, and nanofabrication*, J.Photoch. Photobio. C: Photochemistry Reviews, 13 (**2012**) 28 – 54

[He2005] Q. He, S. Liu, L. Kong, Z. Liu, *A study on the sizes and concentrations of gold nanoparticles by spectra of absorption, resonance Rayleigh scattering and resonance non-linear scattering*, Spectrochim. Acta A 61 (**2005**) 2861-2866

[He2008] C. He, T. Sasaki, Y. Shimizu, N. Koshizaki, *Synthesis of ZnO nanoparticles using nanosecond pulsed laser ablation in aqueous media and their self-assembly towards spindle-like ZnO aggregates*, Appl. Surf. Sci. 254 (**2008**) 2196–2202

[He2013]   Q. He, D. Joy, D. Keffer, *Impact of oxidation on nanoparticle adhesion to carbon substrates*, RSC Adv. 3 (**2013**) 15792-15804

[Heise2011]   G. Heise, M. Dickmann, M. Domke, A. Heiss, T. Kuznicki, J. Palm, I. Richter, H. Vogt, H. Huber, *Investigation of the ablation of zinc oxide thin films on copper–indium-selenide layers by ps laser pulses*, Appl. Phys. A 104 (**2011**) 387–393

[Hernández Battez 2008]   A. Hernández Battez, R. González, J. Viesca, J. Fernández, J. Fernández, A. Machado, R. Chou, J. Riba, *CuO, ZrO₂ and ZnO nanoparticles as antiwear additive in oil lubricants*, Wear 265 (**2008**) 422–428

[Higashi2013]   Y. Higashi, T. Tsuji, M. Tsuji, H. Fujiwara, Y. Ishikawa, N. Koshizaki, *Fabrication of Spherical-Shaped Submicron Particles of ZnO Using Laser-induced Melting of Submicron-sized Source Materials*, WPE-12, (**2013**) Conference on Lasers and Electro-Optics Pacific Rim (CLEO-PR)

[Hiramatsu1998]   M. Hiramatsu, K. Imaeda, N. Horio, M. Nawata, *Transparent conducting ZnO thin films prepared by XeCl excimer laser ablation*, J. Vac. Sci. Technol. A 16 (**1998**) 669-673

[Hong2013]   S. Hong, J. Yeo, G. Kim, D. Kim, H. Lee, J. Kwon, H. Lee, P. Lee, S. Ko, *Nonvacuum, Maskless Fabrication of a Flexible Metal Grid Transparent Conductor by Low-Temperature Selective Laser Sintering of Nanoparticle Ink*, ACS Nano 7 (**2013**) 5024-5031

[Hu2012]   X. Hu, H. Gong, Y. Wang, Q. Chen, J. Zhang, S. Zheng, S. Yang, B. Cao, *Laser-induced reshaping of particles aiming at energy-saving applications*, J. Mater. Chem. 22 (**2012**) 15947-15952

[Huang2001]   M. Huang, S. Mao, H. Feick, H. Yan, Y. Wu, H. Kind, E. Weber, R. Russo, P. Yang, *Room-Temperature*

Ultraviolet Nanowire Nanolasers, Science 292 (**2001**) 1897-1899

[Huang2008]   J. Huang, T. Akita, J. Faye, T. Fujitani, T. Takei, M. Haruta, *Propene Epoxidation with Dioxygen Catalyzed by Gold Clusters*, Angew. Chem. Int. Ed., **2009**, 48, 7862 −78667

[Huang2013]   S. Huang, P. Liu, A. Mokasdar, L. Hou, *Additive manufacturing and its societal impact: a literature review*, Int. J. Adv. Manuf. Technol. 67 (**2013**) 1191-1203

[Hughes2005]   M. Hughes, Y.-J. Xu, P. Jenkins, P. McMorn, P. Landon, D. Enache, A. Carley, G. Attard, G. Hutchings, F. King, E. Stitt, P. Johnston, K. Griffin, C. Kiely, *Tunable gold catalysts for selective hydrocarbon oxidation under mild conditions*, Nature 437 (**2005**) 1132

[Ibrahimkutty2011] S. Ibrahimkutty, J. Kim, M. Cammarata, F. Ewald, J. Choi, H. Ihee, A. Plech, *Ultrafast structural dynamics of the photocleavage of protein hybrid nanoparticles*, ACS Nano 5 (**2011**) 3788-3794

[Ibrahimkutty2012] S. Ibrahimkutty, P. Wagener, A. Menzel, A. Plech, S. Barcikowski, *Nanoparticle formation in a cavitation bubble after pulsed laser ablation in liquid studied with high time resolution small angle x-ray scattering*, Appl. Phys. Lett. 101 (**2012**) 103104

[Inasawa2005]   S. Inasawa, M. Sugiyama, Y. Yamaguchi, *Laser-Induced Shape Transformation of Gold Nanoparticles below the Melting Point: The Effect of Surface Melting*, J. Phys. Chem. B 109 (**2005**) 3104-3111

[Intartaglia2013] R. Intartaglia, G. Das, K. Bagga, A. Gopalakrishnan, A. Genovese, M. Povia, E. Di Fabrizio, R. Cingolani, A. Diaspro, F. Brandi, *Laser synthesis of ligand-free bimetallic nanoparticles for plasmonic applications*, Phys. Chem. Chem. Phys. 15 (**2013**) 3075-3082

[Intartaglia2014]   R. Intartagila, K. Bagga, F. Brandi, *Study on the productivity of silicon nanoparticles by picosecond laser ablation in water: towards gram per hour yield*, Opt. express 22 (**2014**), 3117-3127

[Intartaglia2015]   R. Intartaglia, S. Beke, M. Moretti, F. De Angelis, A. Diaspro, *Fast and cost-effective fabrication of large-area plasmonic transparent biosensor array*, Lab Chip 15 (**2015**) 1343-1349

[Ishikawa2007]   Y. Ishikawa, Y. Shimizu, T. Sasaki, N. Koshizaki, *Boron carbide spherical particles encapsulated in graphite prepared by pulsed laser irradiation of boron in liquid medium*, Appl. Phys. Lett. 91 (**2007**) 161110

[Ishikawa2010]   Y. Ishikawa, Q. Feng, N. Koshizaki, *Growth fusion of submicron spherical boron carbide particles by repetitive pulsed laser irradiation in liquid media*, Appl. Phys. A 99 (**2010**) 797–803

[Ivanova2010]   O. Ivanova, F. Zamborini, *Electrochemical size discrimination of gold nanoparticles attached to glass/indium-tin-oxide electrodes by oxidation in bromide-containing electrolyte*, Anal. Chem. 82 (**2010**) 5844-5850

[Jadraque2008]   M. Jadraque, C. Domingo, M. Martin, *Laser induced effects on ZnO targets upon ablation at 266 and 308nm wavelengths*, J. Appl. Phys. 104 (**2008**) 024306

[Jakobi2011]   J. Jakobi, A. Menéndez-Manjón, V. Chakravadhanula, L. Kienle, P. Wagener, S. Barcikowski, *Stoichiometry of alloy nanoparticles from laser ablation of PtIr in acetone and their electrophoretic deposition on PtIr electrodes*, Nanotechnology 22 (**2011**) 145601 (7pp)

[Jang2004]   D. Jang, B. Oh, D. Kim, *Visualization of microparticle explosion and flow field in nanoparticle synthesis by pulsed laser ablation*, Appl. Phys. A 79 (**2004**) 1149-1151

[Janotti2008]   A. Janotti, C. Van de Walle, *Fundamentals of zinc oxide as a semiconductor*, Rep. Prog. Phys. 72 (**2009**) 126501 (29pp)

[Jeon2007]   H. Jeon, T. Sugiyama, H. Masuhara, T. Asahi, *Study on Electrophoretic Deposition of Size-Controlled Quinacridone Nanoparticles*, J. Phys. Chem. C 111 (**2007**) 14658-14663

[Kabashin2003]   A. Kabashin, M. Meunier, *Synthesis of colloidal nanoparticles during femtosecond laser ablation of gold in water*, J. Appl. Phys. 94 (**2003**) 7941-7943

[Kapoor2000]   S. Kapoor, D. Palit, *Laser-induced fragmentation and melting of cadmium and copper nanoparticles*, Mater. Res. Bull. 35 (**2000**) 2071–2079

[Katayama2014]   T. Katayama, K. Setoura, D. Werner, H. Miyasaka, S. Hashimoto, *Picosecond-to-Nanosecond Dynamics of Plasmonic Nanobubbles from Pump–Probe Spectral Measurements of Aqueous Colloidal Gold Nanoparticles*, Langmuir 30 (**2014**) 9504–9513

[Kathuria1999]   Y. Kathuria, *Microstructuring by selective laser sintering of metallic powder*, Surf. Coat. Tech. 116 (**1999**) 643-647

[Kawaguchi2006]   K. Kawaguchi, J. Jaworski, Y. Ishikawa, T. Sasaki, N. Koshizaki, *Preparation of Gold/Iron Oxide Composite Nanoparticles by a Laser-Soldering Method*, IEEE Trans. Magn. 42 (**2006**) 3620-3622

[Kawaguchi2007]   K. Kawaguchi, J. Jaworski, Y. Ishikawa, T. Sasaki, N. Koshizaki, *Preparation of gold/iron-oxide composite nanoparticles by a unique laser process in water*, J. Magn. Magn. Mater. 310 (**2007**) 2369-2371

[Kawasaki2006]   M. Kawasaki, N. Nishimura, *1064-nm laser fragmentation of thin Au and Ag flakes in acetone for*

                           *highly productive pathway to stable metal
                           nanoparticles*, Appl. Surf. Sci. 253 (**2006**) 2208–2216

[Kelchtermans2013]   A. Kelchtermans, K. Elen, K. Schellens, B. Conings, H.
                           Damm, H. Boyen, J. D'Haen, P. Adriaensens, A. Hardy,
                           M. Van Bael, *Relation between synthesis conditions,
                           dopant position and charge carriers in aluminium-
                           doped ZnO nanoparticles*, RSC Adv. 3 (**2013**) 15254–
                           15262

[Kotaidis2005]   V. Kotaidis, A. Plech, *Cavitation dynamics on the
                           nanoscale,* Appl. Phys. Lett. 87 (**2005**) 213102 2 pp

[Kruth2003]   J. Kruth, X. Wang, L. Froyen, Lasers and materials in
                           selective laser sintering, Assem. Autom. 23 (2003) 357-
                           371

[Kubelka1931]   P. Kubelka, F. Munk, *Ein Beitrag zur Optik der
                           Farbanstriche,* Z. Tech. Phys. 12 (**1931**) 593-601

[Kubelka1948]   P. Kubelka, *New contribution to the optics of intensely
                           light-scattering materials*, J. Opt. Soc. Am. 38 (**1948**)
                           448-457

[Kumar2010]   B. Kumar, R. Thareja, *Synthesis of nanoparticles in laser
                           ablation of aluminum in liquid*, J. Appl. Phys. 108
                           (**2010**) 064906 6pp

[Kuzmin2014]   P. Kuzmin, G. Shafeev, A. Serkov, N. Kirichenko, M.
                           Shcherbina, *Laser-assisted fragmentation of Al
                           particles suspended in liquid*, Appl. Surf. Sci. 294 (**2014**)
                           15-19

[Lau2014a]   M. Lau, S. Barcikowski, *Quantification of mass-specific
                           laser energy input converted into particle properties
                           during picosecond pulsed laser fragmentation of zinc
                           oxide and boron carbide in liquids*,
                           Appl. Surf. Sci. 348 (**2015**) 22-29

[Lau2014b]   M. Lau, S. Barcikowski: *Verfahren zur Herstellung
                           reiner, insbesondere kohlenstofffreier Nanopartikel.*

(2014).   -   Europaeische   Patentanmeldung
EP2735390A1

[Lau2014c]   M. Lau, S. Barcikowski, *Method for manufacture of pure, carbon free nanoparticles*. (2014). - US Patent Application   Publication   US2014/0171523A1, 19.06.2014

[Lau2014d]   M. Lau, I. Haxhiaj, P. Wagener, R. Intartaglia, F. Brandi, J. Nakamura, S. Barcikowski, *Ligand-free gold atom clusters adsorbed on graphene nano sheets generated by oxidative laser fragmentation in water*, Chem. Phys. Lett. 610-611 (2014) 256-260

[Lau2014e]   M. Lau, R. Niemann, M. Bartsch, W. O'Neill, S. Barcikowski, *Near-field-enhanced, off-resonant laser sintering of semiconductor particles for additive manufacturing of dispersed Au–ZnO-micro/nano hybrid structures*, Appl. Phys. A, 114 (2014) 1023-1030

[Lau2015]   M. Lau, A. Ziefuss, T. Komossa, S. Barcikowski, *Inclusion of supported gold nanoparticles into their semiconductor support*, Phys. Chem. Chem. Phys. 17 (2015) 29311-29318

[Li2011a]   X. Li, J. Liu, X. Wang, M. Gao, Z. Wang, X. Zeng, *Preparation of silver spheres by selective laser heating in silver-containing precursor solution*, Opt. Express 19 (2011) 2846

[Li2011]   P. Li, Z. Wei, T. Wu, Q. Peng, Y. Li, *Au-ZnO Hybrid Nanopyramids and Their Photocatalytic Properties*, J. Am. Chem. Soc. 133 (2011) 5660-5663

[Li2012]   X. Li, Y. Shimizu, A. Pyatenko, H. Wang, N, Koshizaki, *Tetragonal zirconia spheres fabricated by carbon-assisted selective laser heating in a liquid medium*, Nanotechnology 23 (2012) 115602 (8pp)

[Li2011b]   X. Li, A. Pyatenko, Y. Shimizu, H. Wang, K. Koga, N. Koshizaki, *Fabrication of Crystalline Silicon Spheres by Selective Laser Heating in Liquid Medium*, Langmuir 27 (**2011**) 5076–5080

[Lica2004]   G. Lica, B. Zelakiewicz, M. Constantinescu, Y. Tong, *Charge Dependence of Surface Plasma Resonance on 2 nm Octanethiol-Protected Au Nanoparticles: Evidence of a Free-Electron System*, J. Phys. Chem. B 108 (**2004**) 19896-19900

[Lide2009]   D. Lide, In CRC *Handbook of chemistry and physics* CRC Press, 90th Edition **2009**, pp 5-80-5-83

[Lin2001]   B. Lin, Z. Fu, Y. Jia, *Green luminescent center in undoped zinc oxide films deposited on silicon substrates*, Appl. Phys. Lett. 79 (**2001**) 943-945

[Lin2003]   Z. Lin, L. Zhigilei, *Thermal excitation of d band electrons in Au: implications for laser-induced phase transformations, High-Power Laser Ablation VI*, edited by C. R. Phipps, Proc. SPIE 6261, 62610U, **2006**.

[Lin2009a]   C.-A. Lin, C.-H. Lee, J.-T. Hsieh, H.-H. Wang, J. Li, J.-L. Shen, W.-H. Chan, H.-I. Yeh, W. Chang, *Synthesis of Fluorescent Metallic Nanoclusters toward Biomedical Application: Recent Progress and Present Challenges*, J. Med. Biol. Eng. 29 (**2009**) 276-283

[Lin2009b]   C. Lin, T. Yang, C. Lee, S. Huang, R. Sperling, M. Zanella, J. Li, J. Shen, H. Wang, H. Yeh, W. Parak, W. Chang, *Synthesis, Characterization, and Bioconjugation of Fluorescent Gold Nanoclusters toward Biological Labeling Applications*, ACS Nano 3 (**2009**) 395-401

[Link1999a]   S. Link, M. El-Sayed, *Spectral Properties and Relaxation Dynamics of Surface Plasmon Electronic Oscillations in Gold and Silver Nanodots and Nanorods*, J. Phys. Chem. B 103 (**1999**) 8410-8426

[Link1999b]   S. Link, M. El-Sayed, *Size and Temperature Dependence of the Plasmon Absorption of Colloidal Gold Nanoparticles*, J. Phys. Chem. B 103 (**1999**) 4212-4217

[Link1999c]   S. Link, C. Burda, M. B. Mohamed, B. Nikoobakht, and M. A. El-Sayed, *Laser Photothermal Melting and Fragmentation of Gold Nanorods: Energy and Laser Pulse-Width Dependence*, J. Phys. Chem. A, 103 (**1999**) 1165-1170

[Liu2015]   D. Liu, C. Li, F. Zhou, T. Zhang, H. Zhang, X. Li, G. Duan, W. Cai, Y. Li, *Rapid Synthesis of Monodisperse Au Nanospheres through a Laser Irradiation –Induced Shape Conversion, Self-Assembly and Their Electromagnetic Coupling SERS Enhancement*, Scientific Reports 5 (**2015**) 7686

[Luo2012]   L. Luo, M. Rossell, D. Xie, R. Erni, M. Niederberger, *Microwave-Assisted Nonaqueous Sol–Gel Synthesis: From Al:ZnO Nanoparticles to Transparent Conducting Films*, ACS Sustain. Chem. Eng. 1 (**2013**) 152–160

[Mafuné2001]   F. Mafuné, J. Kohno, Y. Takeda, T. Kondow, H. Sawabe, *Formation of Gold Nanoparticles by Laser Ablation in Aqueous Solution of Surfactant*, J. Phys. Chem. B 105 (**2001**) 5114-5120

[Mafuné2002]   F. Mafuné, J. Kohno, Y. Takeda, T. Kondow, *Growth of Gold Clusters into Nanoparticles in a Solution Following Laser-Induced Fragmentation*, J. Phys. Chem. B 106 (**2002**) 8555-8561

[Mafuné2004]   F. Mafuné, T. Kondow, *Selective laser fabrication of small nanoparticles and nano-networks in solution by irradiation of UV pulsed laser onto platinum nanoparticles*, Chem. Phys. Lett. 383 (**2004**) 343–347

[Mafuné2014]   F. Mafuné, T. Okamoto, M. Ito, *Surfactant-free small Ni nanoparticles trapped on silica nanoparticles prepared by pulsed laser ablation in liquid*, Chem. Phys. Lett. 591 (**2014**) 193-196

[Maiman1960a]   T. Maiman, *Stimualted Optical Radiation in Ruby*, Nature 187 (**1960**), 493 - 494

[Maiman1960b]   T. Maiman, *Optical and Microwave-Optical Experiments in Ruby*, Phys. Rev. Lett. 4 (**1960**) 564

[Maiman1962]   T. Maiman, *Solid state laser and iraser studies*, Solid-State Electronics 4 (**1962**) 236-249

[Manshina2015]   A. Manshina, A. Povolotskiy, A. Povolotckaia, A. Kireev, Y. Petrov, S. Tunik, *Annealing effect: Controlled modification of the structure, composition and plasmon resonance of hybrid Au–Ag/C nanostructures*, Appl. Surf. Sci. 353 (**2015**) 11-16

[Marzun2014]   G. Marzun, C. Streich, S. Jendrzej, S. Barcikowski, P. Wagener, *Adsorption of Colloidal Platinum Nanoparticles to Supports: Charge Transfer and Effects of Electrostatic and Steric Interactions*, Langmuir 30 (**2014**) 11928–11936

[Marzun2015]   G. Marzun, J. Nakamura, X. Zhang, S. Barcikowski, P. Wagener, *Size control and supporting of palladium nanoparticles made by laser ablation in saline solution as a facile route to heterogeneous catalysts*, Appl. Surf. Sci. 348, (**2015**) 75-84

[McCarthy1934]   E. McCarthy, *Mariotte's bottle*, Science 27 (**1934**) 100

[Meier2014]   J. Meier, C. Galeano, I. Katsounaros, J. Witte, H. Bongard, A. Topalov, C. Baldizzone, S. Mezavilla, F. Schüth, K. Mayrhofer, *Design criteria for stable Pt/C fuel cell catalysts*, Beilstein J. Nanotechnol. 5 (**2014**) 44-67

[Menendez-Manjon2007] A. Menendez-Manjon, S. Barcikowski, *Hydrodynamic size distribution of gold nanoparticles controlled by repetition rate during pulsed laser ablation in water*, Appl. Surf. Sci. 257 (**2011**) 4285–4290

[Merk2014]   V. Merk, C. Rehbock, F. Becker, U. Hagemann, H. Nienhaus, S. Barcikowski, *In situ non-DLVO stabilization of surfactant-free, plasmonic gold nanoparticles: Effect of Hofmeister's anions*, Langmuir 30 (**2014**) 4213-4222

[Messina2013]   G. Messina, P. Wagener, R. Streubel, A. De Giacomo, A. Santagata, G. Compagnini, S. Barcikowski, *Pulsed laser ablation of a continuously-fed wire in liquid flow for high-yield production of silver nanoparticles*, Phys. Chem. Chem. Phys. 15 (**2013**) 3093-3098

[Mie1908]   G. Mie, *Beiträge zur Optik trüber Medien, speziell kolloidaler Metallösungen*, Ann. Phys.4 (**1908**) 376-445

[Morkoc2009]   H. Morkoc, Ü. Özgür, *Zinc Oxide Fundamentals, Materials and Device Technology*, Wiley-VCH Verlag GmbH & Co. KGa, 2009, ISBN:978-3-527-40213-9

[Murphy2008]   C. Murphy, A. Gole, J. Stone, P. Sisco, A. Alkilany, E. Goldsmith, S. Baxter, *Gold Nanoparticles, in Biology: Beyond Toxicity to Cellular Imaging*, Acc. Chem. Res. 41 (**2008**) 1721-1730

[Murphy2008a]   C. Murphy, *Sustainability as an emerging design criterion in nanoparticle synthesis and applications*, J. Mater. Chem. 18 (**2008**) 2173-2176

[Murr2012]   L. Murr, S. Gaytan, D. Ramirez, E. Martinez, J. Hernandez, K. Amato, P. Shindo, F. Medina, R. Wicker, *Metal Fabrication by Additive Manufacturing Using Laser and Electron Beam Melting Technologies*, J. Mater. Sci. Technol. 28 (**2012**) 1-14

[Muto2007]   H. Muto, K. Yamada, K. Miyajima, F. Mafuné, *Estimation of surface oxide on surfactant-free gold nanoparticles laser-ablated in water*, J. Phys. Chem. C 111 (**2007**) 17221-17226

[Na2010]   S. Na, C. Park, *First-Principples Study of the Surface Energy and the Atom Cohesion of Wurzite ZnO and ZnS – Implications for Nanostructure Formation*, J. Kor. Phys. Soc. 56 (**2010**) 498-502

[Nachev2012]   P. Nachev, D. Zand, V. Coger, P. Wagener, K. Reimers, P. Vogt, S. Barcikowski, A. Pich, *Synthesis of hybrid microgels by coupling of laser ablation and polymerization in aqueous medium*, J. Laser Appl. 24 (**2012**) 042012 7pp

[Nakamura2014]   T. Nakamura, Z. Yuan, S. Adachi, *Micronization of red-emitting $K_2SiF_6:Mn^{4+}$ phosphor by pulsed laser irradiation in liquid*, Appl. Surf. Sci. 320 (**2014**) 514-518

[Nakamura2015]   M. Nakamura, A. Oyane, I. Sakamaki, Y. Ishikawa, Y. Shimizu, K. Kawaguchi, *Laser-assisted one-pot fabrication of calcium phosphate-based submicrospheres with internally crystallized magnetite nanoparticles through chemical precipitation*, Phys. Chem. Chem. Phys. 17 (**2015**) 8836−8842

[Neddersen1993]   J. Neddersen, G. Chumanov, T. Cotton, *Laser Ablation of Metals: A New Method for Preparing SERS Active Colloids*, Appl. Spectrosc. 47 (**1993**) 1959-1964

[Neumeister2014]   A. Neumeister, J. Jakobi, C. Rehbock, J. Moysig, S. Barcikowski, *Monophasic ligand-free alloy nanoparticle synthesis determinants during pulsed laser ablation of bulk alloy and consolidated microparticles in water*, Phys. Chem. Chem. Phys. 16 (**2014**) 23671-23678

[Nichols2006]   W. Nichols, T. Kodaira, Y. Sasaki, Y. Shimizu, T. Sasaki, N. Koshizaki, *Zeolite LTA Nanoparticles Prepared by Laser-Induced Fracture of Zeolite Microcrystals*, J. Phys. Chem. B 110 (**2006**) 83-89

[Nijhuis2006]   T. Nijhuis, B. Weckhuysen, *The direct epoxidation of propene over gold-titania catalysts – A study into the* kinetic mechanism and deactivation, Catal. Today 117 (**2006**) 84-89

[Oba2008]   F. Oba, A. Togo, I. Tanaka, *Defect energetics in ZnO: A hybrid Hartree-Fock density functional study*, Phys. Rev. B 77 (**2008**) 245202

[Özgür2005]   Ü. Özgür, Y. Alivov, C. Liu, A. Teke, M. Reshchikov, S. Doğan, V. Avrutin, S.-J. Cho, H. Morkç , *A comprehensive review of ZnO materials and devices*, J. Appl. Phys. 98 (**2005**) 41301

[Palpant1998]   B. Palpant, B. Prével, J. Lermé, E. Cottancin, M. Pellarin, M. Treilleux, A. Perez, J. Vialle, M. Broyer, *Optical properties of gold clusters in the size range 2–4 nm*, Phys. Rev. B 57 (**1998**) 1963

[Pan2007]   Y. Pan, S. Neuss, A. Leifert, M. Fischler, F. Wen, U. Simon, G. Schmid, W. Brandau, W. Jahnen-Dechent, *Size-Dependent Cytotoxicity of Gold Nanoparticles*, small 3 (**2007**) 1941 – 1949

[Pestryakov2004]   A. Pestryakov, V. Petranovskii, A. Kryazhov, O. Ozhereliev, N. Pfänder, A. Knop-Gericke, *Study of copper nanoparticles formation on supports of different nature by UV–Vis diffuse reflectance spectroscopy*, Chem. Phys. Lett. 385 (**2004**) 173-176

[Petersen2011]   S. Petersen, A. Barchanski, U. Taylor, S. Klein, D. Rath, S. Barcikowski, *Penetratin-Conjugated Gold Nanoparticles - Design of Cell-Penetrating Nanomarkers by Femtosecond Laser Ablation*, J. Phys. Chem. C 115 (**2011**) 5152–5159

[Pfeiffer2014]   C. Pfeiffer, C. Rehbock, D. Hühn, C. Carrillo-Carrion, D. de Aberasturi, V. Merk, S. Barcikowski, W. Parak, *Interaction of colloidal nanoparticles with their local environment: the (ionic) nanoenvironment around nanoparticles is different from bulk and determines the*

                          physico-chemical properties of the nanoparticles, J. R.
                          Soc. Interface 11 (**2014**) 20130931

[Philip2012]   R. Philip, P. Chantharasupawong, H. Qian, R. Jin, J.
                          Thomas, *Evolution of Nonlinear Optical Properties:
                          From Gold Atomic Clusters to Plasmonic Nanocrystals*,
                          Nano Lett. 12 (**2012**) 4661–4667

[Pyatenko2013]   A. Pyatenko, H. Wang, N. Koshizaki, T. Tsuji,
                          *Mechanism of pulse laser interaction with colloidal
                          nanoparticles*, Laser Photon. Rev. 7 (**2013**) 596-604

[Qian2011]   H. Qian, M. Zhu, Z. Wu, R. Jin, *Quantum sized gold
                          nanoclusters with atomic precision*, Acc. Chem. Res. 45
                          (**2011**) 1470-1479

[Raab2011]   C. Raab, M. Simkó, U. Fiedeler, M. Nentwich, A. Gazsó,
                          Production of nanoparticles and nanomaterials, nano
                          trust dossiers 6 (**2011**) 4 pp

[Rajeswari2011]   N. Rajeswari Yogamalar, A. Chandra Bose, *Absorption–
                          emission study of hydrothermally grown Al:ZnO
                          nanostructures*, J. Alloys Comp. 509 (**2011**) 8493–8500

[Rance2010]   G. Rance, D. Marsh, S. Bourne, T. Reade, A. Khlobystov,
                          *Van der waals interactions between nanotubes and
                          nanoparticles for controlled assembly of composite
                          nanostructures*, ACS nano 4 (**2010**) 4920-4928

[Rayleigh1892]   J. W. S., 3rd Baron Rayleigh, *On the instability of
                          cylindrical fluid surfaces*, Phil. Mag. (5) 34 (**1892**) 177-
                          180

[Rayleigh1899]   J. W. S., 3rd Baron Rayleigh, *On the Calculation of the
                          Frequency of Vibration of a System in Its Gravest Mode,
                          With an Example from Hydrodynamics*, Phil. Mag. (5)
                          47 (**1899**) p. 379.

[Rayleigh1910]   J. W. S., 3rd Baron Rayleigh, *Colours of the sea and sky*,
                          Nature 83 (**1910**) 48-50

[Rehbock2013]   C. Rehbock, V. Merk, L. Gamrad, R. Streubel, S. Barcikowski, *Size control of laser-fabricated surfactant-free gold nanoparticles with highly diluted electrolytes and their subsequent bioconjugation*, Phys. Chem. Chem. Phys. 15 (**2013**) 3057-3067

[Rehbock2014]   C. Rehbock, J. Jakobi, L. Gamrad, S. v.d. Meer, D. Tiedemann, U. Taylor, W. Kues, D. Rath, S. Barcikowski, *Current state of laser synthesis of metal and alloy nanoparticles as ligand-free reference materials for nano-toxicological assays*, Beilstein J. Nanotechnol. 5 (**2014**) 1523-1541

[Rehbock2014a]   C. Rehbock, J. Zwartscholten, S. Barikowski, *Biocompatible Gold Submicrometer Spheres with Variable Surface Texture Fabricated by Pulsed Laser Melting in Liquid*, Chem. Lett. 43 (2014) 1502-1504

[Riabinina2011]   D. Riabinina, J. Zhang, M. Chaker, J. Margot, D. Ma, P. Tijssen, *Control of plasmon resonance of gold nanoparticles via excimer laser irradiation*, Appl. Phys. A 102 (**2011**) 153–160

[Risch2011]   A. Risch, R. Hellmann, *Laser scribing of gallium doped zinc oxide thin films using picosecond laser*, Appl. Surf. Sci. 258 (**2011**) 1849-1853

[Sajti2010]   L. Sajti, R. Sattari, B. Chichkov, S. Barcikowski, *Gram Scale Synthesis of Pure Ceramic Nanoparticles by Laser Ablation in Liquid*, J. Phys. Chem. C 114 (**2010**) 2421–2427

[Sajti2011]   C. Sajti, A. Barchanski, P. Wagener, S. Klein, S. Barcikowski, *Delay Time and Concentration Effects During Bioconjugation of Nanosecond Laser-Generated Nanoparticles in a Liquid Flow*, J. Phys. Chem. C 115 (**2011**) 5094–5101

[Sakurai1997]   H. Sakurai, A.Ueda, T. Kobayshi, M. Haruta, *Low-temperature water–gas shift reaction over gold deposited on TiO₂*, Chem. Commun. (**1997**) 271-272

[Schade2014]   L. Schade, S. Franzka, K. Dzialkowski, S. Hardt, H. Wiggers, S. Reichenberger,P. Wagener, N. Hartmann, *Resonant photothermal laser processing of hybrid gold/titania nanoparticle films*, Appl. Surf. Sci. 336 (**2014**) 48-52

[Schaumberg2014] C. Schaumberg, M. Wollgarten, K. Rademann, *Metallic Copper Colloids by Reductive Laser Ablation of Nonmetallic Copper Precursor Suspensions*, J. Phys. Chem. A 118 (**2014**) 8329–8337

[Schindler1964]   P. Schindler. H. Althaus, W. Feitknecht, *Löslichkeitsprodukte und Freie Bildungsenthalpien von Zinkoxid, amorphem Zinkhydroxid, β₁, β₂, γ-, δ- und ε-Zinkhydroxid*, Helv. Chim. Acta 47 (**1964**) 982-991

[Schindler1965]   P. Schindler, H. Althaus, F. Hofer, W. Minder, *Loeslichkeitsprodukte von Zinkoxid, Kupferhydroxid und Kupferoxid in Abhängigkeit von Teilchengrösse und molarer Oberflache. Ein Beitrag zur Thermodynamik von Grenzflachen fest-flussig*, Helv. Chim. Acta 48 (**1965**) 1204-1215

[Schmid1981]   G. Schmid, R. Pfeil, R. Boese, F. Bandermann, S. Meyer, G. Calis, J. Velden, *Au₅₅[P(C₆H₅)₃]₁₂Cl₆ – ein Goldcluster ungewöhnlicher Größe*, Chem. Ber. 114 (**1981**) 3634-3642

[Schmid2008]   G. Schmid, *The relevance of shape and size of Au55 clusters*, Chem. Soc. Rev. 37 (**2008**) 1909-1930

[Shang2011]   L. Shang, S. Dong, G. Nienhaus, *Ultra-small fluorescent metal nanoclusters: Synthesis and biological applications*, Nano Today 6 (**2011**) 401-418

[Sibbett2012]   W. Sibbett, A. Lagatsky, C. Brown, *The development and application of femtosecond laser systems*, Opt. Express 20 (**2012**) 6989-7001

[Siburian2012]   R. Siburian, J. Nakamura, *Formation process of Pt subnano-clusters on graphene nanosheets*, J. Phys. Chem. C 116 (**2012**) 22947-22953

[Siburian2013]   R. Siburian, T. Kondo, J. Nakamura, *Size Control to a Sub-Nanometer Scale in Platinum Catalysts on Graphene*, J. Phys. Chem. C 117 (**2013**) 3635-3645

[Singh2011]   M. Singh, A. Agarwal, R. Goapl, R. Swarnkar, R. Kotnala, *Dumbbell shaped nickel nanocrystals synthesized by a laser induced fragmentation method*, J. Mater. Chem. 21 (**2011**) 11074

[Sowa-Soehle2013] E. Sowa-Soehle, A. Schwenke, P. Wagener, A.Weiss, H. Wiegel, C. Sajti, A. Haverich, S. Barcikowski, A. Loos, *Antimicrobial efficacy, cytotoxicity, and ion release of mixed metal (Ag, Cu, Zn, Mg) nanoparticle polymer composite implant material*, BioNanoMat 14 (**2013**) 217–227

[Spinelli2012]   P. Spinelli, A. Polman, *Prospects of near-field plasmonic absorption enhancement in semiconductor materials using embedded Ag nanoparticles*, Opt. Express 20 (**2012**) A641-A654

[Srikant1998]   V. Srikant, D. Clarke, *On the optical band gap of zinc oxide*, J. Appl. Phys. 83 (**1998**) 5447-5451

[Stamatakis1990] P. Stamatakis, B. Palmer, G. Salzmann, C. Bohren, T. Allen, *Optimum particle size of titanium dioxide and zinc oxide for attenuation of ultraviolet radiation*, J. Coat. Technol. 62 (**1990**) 95-98

[Steen2010]   W. Steen, J. Mazumder, Laser Material Processing, Springer London Dordrecht Heidelberg New York 4th Edition (**2010**) ISBN 978-84996-061-8

[Strasser2014]   M. Strasser, K. Setoura, U. Langbein, S. Hashimoto, *Computational Modeling of Pulsed Laser-Induced Heating and Evaporation of Gold Nanoparticles*, J. Phys. Chem. C 118 (**2014**) 25748–25755

[Stratakis2012]   M. Stratakis, H. Garcia, *Catalysis by Supported Gold Nanoparticles: Beyond Aerobic Oxidative Processes*, Chem. Rev. 112 (**2012**) 4469-4506

[Strunk2009]   J. Strunk, K. Kähler, X. Xia, M. Comotti, F. Schüth, T. Reinecke, M. Muhler, *Au/ZnO as catalyst for methanol synthsis: The role of oxygen vacancies*, Appl. Cat. A 359 (**2009**) 121-128

[Sugiyama2006]   T. Sugiyma, T. Asahi, H. Takeuchi, H. Masuhara, *Size and Phase Control in Quinacridone Nanoparticle Formation by Laser Ablation in Water*, Jpn. J. Appl. Phys. 45 (**2006**) 384–388

[Sugiyama2011]   T. Sugiyama, T. Asahi, *Fabrication of the Smallest Organic Nanocolloids by a Top-down Method Based on Laser Ablation*, Chem. Rec. 11 (**2011**) 54-58

[Swiatkowska-Warkocka2012] Z. Swiatkowska-Warkocka, K. Kawaguchi, Y. Shimizu, A. Pyatenko, H. Wang, N. Koshizaki, *Synthesis of Au-Based Porous Magnetic Spheres by Selective Laser Heating in Liquid*, Langmuir 28 (**2012**) 4903-4907

[Swiatkowska-Warkocka2013] Z. Swiatkowska-Warkocka, K. Koga, K. Kawaguchi, H. Wang, A. Pyatenko, N. Koshizaki, *Pulsed laser irradiation of colloidal nanoparticles: a new synthesis route for the production of non-equilibrium bimetallic alloy submicrometer spheres*, RSC Adv. 3 (**2013**) 79–83

[Sylvestre2004]   J.-P. Sylvestre, S. Poulin, A. Kabashin, E. Sacher, M. Meunier, J. Luong, *Surface Chemistry of Gold Nanoparticles Produced by Laser Ablation in Aqueous Media*, J. Phys. Chem. B 108 (**2004**) 16864-16869

[Sylvestre2004a]   J.-P. Sylvestre, A. Kabashin, E. Sacher, M. Meunier, J. Luong, *Stabilization and Size Control of Gold Nanoparticles during Laser Ablation in Aqueous Cyclodextrins*, J. Am. Chem. Soc. 126 (**2004**) 7176-7177

[Sylvestre2011]   J. Sylvestre, M. Tang, A. Furtos, G. Leclair, M. Meunier, J. Leroux, *Nanonization of megestrol acetate by laser fragmentation in aqueous milieu*, J. Controll. Rel. 149 (**2011**) 273–280

[Taguchi2009]   M. Taguchi, S. Takami, T. Naka, T. Adschiri, *Growth Mechanism and Surface Chemical Characteristics of Dicarboxylic Acid-Modified CeO2 Nanocrystals Produced in Supercritical Water: Tailor-Made Water-Soluble CeO2 Nanocrystals*, Cryst. Growth Des. 9 (**2009**) 5297-5303

[Takeda2014]   Y. Takeda,F. Mafuné, *Formation of wide bandgap cerium oxide nanoparticles by laser ablation in aqueous solution*, Chem. Phys. Lett. 599 (**2014**) 110-115

[Tamaki2000]   Y. Tamaki, T. Asahi, H. Masuhara, *Tailoring nanoparticles of aromatic and dye molecules by excimer laser irradiation*, Appl. Surf. Sci. 168 (**2000**) 85-88

[Tamaki2002]   Y. Tamaki, T. Asahi, H. Masuhara, *Nanoparticle Formation of Vanadyl Phthalocyanine by Laser Ablation of Its Crystalline Powder in a Poor Solvent*, J. Phys. Chem. A 106 (**2002**) 2135-2139

[Tolochko2000]   N. Tolochko, Y. Khlopkov, S. Mozzharov, M. Ignatiev, T. Laoui, V. Titov, *Absorptance of powder materials suitable for laser sintering*, Rapid Prototyping J. 6 (**2001**) 155-161

[Tsuji2013]   T. Tsuji, T. Yahata, M. Yasutomo, K. Igawa, M. Tsuji, Y. Ishikawa, N. Koshizaki, *Preparation and investigation of the formation mechanism of submicron-sized spherical particles of gold using laser ablation and laser irradiation in liquids*, Phys. Chem. Chem. Phys. 15 (**2013**) 3099-3107

[Tsuji2013a]   T. Tsuji, Y. Higashi, M. Tsuji, H. Fujiwara, Y. Ishikawa, N, Koshizaki, *Fabrication of Spherical-Shaped Submicron Particles of ZnO Using Laser-induced Melting of Submicron-sized Source Materials*, J. Laser Micro. Nanoen. 8 (**2013**) 292-295

[Tsuji2015]   T. Tsuji, Y. Higashi, M. Tsuji, Y. Ishikawa, N. Koshizaki, *Preparation of submicron-sized spherical particles of gold using laser-induced melting in liquids and low-toxic stabilizing reagent*, Appl. Surf. Sci. 348 (**2015**) 10-15

[Turkevich1951] J. Turkevich, P. Stevenson, J. Hillier, *A Study of the Nucleation and Growth Process in the Synthesis of Colloidal Gold*, Discuss. Faraday Soc., **1951**

[Turner2008]   M. Turner, V. Golovko, O. Vaughan, P. Abdulkin, A. Berenguer-Murcia, M. Tikhov, B. Johnson, R. Lambert, *Selective oxidation with dioxygen by gold nanoparticle catalysts derived from 55-atom clusters*, Nature 454 (**2008**) 981

[Tyndall1869]   J. Tyndall, *On the blue color of the sky, the polarization of skylight, and polarization of light by cloudy matter generally*, J. Franklin Inst. 88 (**1869**) 34-40

[Ullmann2002]   M. Ullmann, S. Firedlander, A. Schmidt-Ott, Nanoparticle formation by laser ablation, J. Nanopart. Res. 4 (2002) 499-509

[Usui2005]   H. Usui, Y. Shimizu, T. Sasaki, N. Koshizaki, *Photoluminescence of ZnO Nanoparticles Prepared by Laser Ablation in Different Surfactant Solutions*, J. Phys. Chem. B 109 (**2005**) 120-124

[Usui2006]   H. Usui, T. Sasaki, N. Koshizaki, *Optical Transmittance of Indium Tin Oxide Nanoparticles Prepared by Laser-*

Induced Fragmentation in Water, J. Phys. Chem. B 110 (**2006**) 12890-12895

[Vogel19999]   A. Vogel, J. Noack, K. Nahen, D. Theisen, S. Busch, U. Parlitz, D. Hammer, G. Noojin, B. Rockwell, R. Birngruber, *Energy balance of optical breakdown in water at nanosecond to femtosecond time scales*, Appl. Phys. B 68 (**1999**) 271–280

[Wagener2010]   P. Wagener, A. Schwenke, B. Chichkov, S. Barcikowski, *Pulsed Laser Ablation of Zinc in Tetrahydrofuran: Bypassing the Cavitation Bubble*, J. Phys. Chem. C 114 (**2010**) 7618–7625

[Wagener2010a]   P. Wagener, S. Barcikowski, *Laser fragmentation of organic microparticles into colloidal nanoparticles in a free liquid jet*, Appl. Phys. A 101 (**2010**) 435-439

[Wagener2011]   P. Wagener, J. Jakobi, S. Barcikowski, *Organic Nanoparticles Generated by Combination of Laser Fragmentation and Ultrasonication in Liquid*, J. Laser Micro Nanoen. 6 (**2011**) 59-63

[Wagener2012a]   P. Wagener, A. Schwenke, S. Barcikowski, *How Citrate Ligands Affect Nanoparticle Adsorption to Microparticle Supports*, Langmuir 28 (**2012**) 6132–6140

[Wagener2012]   P. Wagener, M. Lau, S. Breitung-Faes, A. Kwade, S. Barcikowski, *Physical fabrication of colloidal ZnO nanoparticles combining wet-grinding and laser fragmentation*, Appl. Phys. A 108 (**2012**) 793–799

[Wagener2013]   P. Wagener, S. Ibrahimkutty, A. Menzel, A. Plech, S. Barcikowski, *Dynamics of silver nanoparticle formation and agglomeration inside the cavitation bubble after pulsed laser ablation in liquid*, Phys. Chem. Chem. Phys. 15 (**2013**) 3068–3074

[Wang2004]   Z. Wang, *Zinc oxide nanostructures: growth, properties and applications*, J. Phys.: Condens. Mater 16 (**2004**) 829-858

[Wang2010]   H. Wang, A. Pyatenko, K. Kawaguchi, X. Li, Z. Swiatkowska-Warkocka, N. Koshizaki, *Selective Pulsed Heating for the Synthesis of Semiconductor and Metal Submicrometer Spheres*, Angew. Chem. Int. Ed. 49 (**2010**) 6361 –6364

[Wang2011]   H. Wang, N. Koshizaki, L. Li, L. Jia, K. Kawaguchi, X. Li, A. Pyatenko, Z. Swiatkowska-Warkocka, Y. Bando, D. Golberg, *Size-Tailored ZnO Submicrometer Spheres: Bottom-Up Construction, Size-Related Optical Extinction, and Selective Aniline Trapping*, Adv. Mater. 23 (**2011**) 1865–1870

[Wang2011a]   H. Wang, M. Miyauchi, Y. Ishikawa, A. Pyatenko, N. Koshizaki, Y. Li, L. Li, X. Li, Y. Bando, D. Golberg, *Single-Crystalline Rutile TiO2 Hollow Spheres: Room-Temperature Synthesis, Tailored Visible-Light-Extinction, and Effective Scattering Layer for Quantum Dot-Sensitized Solar Cells*, J. Am. Chem. Soc. 113 (**2011**) 19102–19109

[Wang2012]   H. Wang, K. Kawaguchi, A. Pyatenko, X. Ki, Z. Swiatkowska-Warkocka, Y. Katou, N. Koshizaki, *General Bottom-Up Construction of Spherical Particles by Pulsed Laser Irradiation of Colloidal Nanoparticles: A Case Study on CuO*, Chem. Eur. J. 18 (**2012**) 163 – 169

[Wang2013]   H. Wang, L. Jia, L. Li, X. Li, Z. Swiatkowska-Warkocka, K. Kawaguchi, A. Pyatenko, N. Koshizaki, *Photomediated assembly of single crystalline silver spherical particles with enhanced electrochemical performance*, J. Mater. Chem. A 1 (**2013**) 692–698

[Wen2011]   F. Wen, Y. Dong, L. Feng, S. Wang, S. Zhang, X. Zhang, *Horseradish Peroxidase Functionalized Fluorescent*

Gold Nanoclusters for Hydrogen Peroxide Sensing, Anal. Chem. 83 (**2011**) 1193–1196

[Werner2011a]   D. Werner, S. Hashimoto, *Improved Working Model for Interpreting the Excitation Wavelength- and Fluence-Dependent Response in Pulsed Laser-Induced Size Reduction of Aqueous Gold Nanoparticles*, J. Phys. Chem. C 115 (**2011**) 5063-5072

[Werner2011]   D. Werner, A. Furube, T. Okamoto, S. Hashimoto, *Femtosecond Laser-Induced Size Reduction of Aqueous Gold Nanoparticles: In Situ and Pump–Probe Spectroscopy Investigations Revealing Coulomb Explosion*, J. Phys. Chem. C 115 (**2011**) 8503-8512

[Werner2013]   D. Werner, S. Hashimoto, *Controlling the Pulsed-Laser-Induced Size Reduction of Au and Ag Nanoparticles via Changes in the External Pressure, Laser Intensity, and Excitation Wavelength*, Langmuir 29 (**2013**) 1295–1302

[Wöll2007]   C. Wöll, *The chemistry and physics of zinc oxide surfaces*, Progress in Surface Science 82 (**2007**) 55–120

[Xue2014]   Y. Xue, X. Li, H. Li, W. Zhang, *Quantifying thiol–gold interactions towards the efficient strength control*, Nat. Commun. (**2014**) 5:4348

[Yamamoto2011]   T. Yamamoto, Y. Shimotsuma, M. Sakakura, M. Nishi, K. Miura,K. Hirao, *Intermetallic Magnetic Nanoparticle Precipitation by Femtosecond Laser Fragmentation in Liquid*, Langmuir 27 (**2011**) 8359–8364

[Yan2010]   Z. Yan, R. Bao, Y. Huang, A. Caruso, S. Qadri, C. Zoica Dinu, D. Chrisey, *Excimer Laser Production, Assembly, Sintering, and Fragmentation of Novel Fullerene-like Permalloy Particles in Liquid*, J. Phys. Chem. C 114 (**2010**) 3869–3873

[Yang2012]   Y. Yang, P. Lan, M. Wang, T. Wie, R. Tan, W. Song, *Nearly full-dense and fine-grained AZO:Y ceramics sintered from the corresponding nanoparticles*, Nanoscale Res. Lett. 7 (**2012**) 481

[Yeh1998]   Y. Yeh, M. Yeh, Y. Lee, C. Yeh, *Formation of Cu Nanoparticles from CuO Powder by Laser Ablation in 2-Propanol*, Chem. Lett. 27 (**1998**) 1183-1184

[Yeh1999]   M. Yeh, Y. Yang, Y. Lee, H. Lee, Y. Yeh, C. Yeh, *Formation and Characteristics of Cu Colloids from CuO Powder by Laser Irradiation in 2-Propanol*, J. Phys. Chem. B 103 (**1999**) 6851-6857

[Yu2012]   W. Yu, M. D. Porosoff, J. G. Chen, *Review of Pt-Based Bimetallic Catalysis: From Model Surfaces to Supported Catalysts*, Chem. Rev. 112 (**2012**) 5780-5817

[Zak2011]   A. Zak, R. Razali, W. Majid, M. Darroudi, *Synthesis and characterization of a narrow size distribution of zinc oxide nanoparticles*, Int. J. Nanomedicine 6 (**2011**) 1399–1403

[Zeng2005]   H. Zeng, W. Cai, Y. Li, J. Hu, P. Liu, *Composition/Structural Evolution and Optical Properties of ZnO/Zn Nanoparticles by Laser Ablation in Liquid Media*, J. Phys. Chem. B 109 (**2005**) 18260-18266

[Zeng2010]   H. Zeng, G. Duan, Y. Li, S. Yang, X. Xu, W. Cai, *A Blue Luminescence of ZnO Nanoparticles Based on Non-Equilibrium Processes: Defect Origins and Emission Controls*, Adv. Funct. Mater. 20 (**2010**) 561–572

[Zeng2011]   H. Zeng, S. Yang, W. Cai, *Reshaping Formation and Luminescence Evolution of ZnO Quantum Dots by Laser-Induced Fragmentation in Liquid*, J. Phys. Chem. C 115 (**2011**) 5038–5043

[Zhang2003]   J. Zhang, J. Worley, S. Dénommée, C. Kingston, Z. Jakubek, Y. Deslandes, M. Post, B. Simard, *Synthesis of Metal Alloy Nanoparticles in Solution by Laser Irradiation of a Metal Powder Suspension*, J. Phys. Chem. B 107 (**2003**) 6920-6923

[Zheng2004]   J. Zheng, C. Zhang, R. Dickson, *Highly Fluorescent, Water-Soluble, Size-Tunable Gold Quantum Dots*, Phys. Rev. Lett. 93 (**2004**) 77402

[Zheng2007]   J. Zheng, P. Nicovich, R. Dickson, *Highly Fluorescent Noble-Metal Quantum Dots*, Annu. Rev. Phys. Chem 58 (**2007**) 409-431

[Zhigilei1998] L. Zhigilei, B. Garrision, *Computer simulation study of damage and ablation of submicron particles from short-pulse laser irradiation*, Appl. Surf. Sci. 127–129 (**1998**) 142–150

[Zhu2007]   B. Zhu, C. Xie, A. Wang, J. Wu, R. Wu, J. Liu, *Laser sintering ZnO thick films for gas sensor application*, J. Mater. Sci. 42 (**2007**) 5416-5420

[Zhukov2012]   V. Zhukov, V. Tyuterev, E. Chulkov, *Electron–phonon relaxation and excited electron distribution in zinc oxide and anatase*, J. Phys.: Condens. Mater 24 (**2012**) 405802 (10p)

8 Appendix

8.1 Supplementary Data

This appendix is divided into nine sections corresponding to the experimental and results sections and gives a comprehensive supporting information to the experimental results described in chapter 4.

List of Figures in the Appendix

Supplementary Data 4.1.1

M. Lau, S. Barcikowski, *Quantification of mass-specific laser energy input converted into particle properties during picosecond pulsed laser fragmentation of zinc oxide and boron carbide in liquids*, Appl. Surf. Sci. 348 (**2015**) 22-29

Quantification of mass-specific laser energy input converted into particle properties during picosecond pulsed laser fragmentation of zinc oxide and boron carbide in liquids

Marcus Lau, Stephan Barcikowski

Technical Chemistry I, University of Duisburg-Essen and Center for Nanointegration Duisburg-Essen (CENIDE), Universitaetsstr. 7, 45141 Essen

Supporting Information

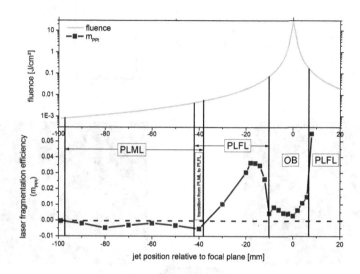

Figure 8-1:   Process windows for pulsed laser melting in liquid (PLML), pulsed laser fragmentation in liquid (PLFL) and the optical breakdown (OB) for the fluences created by 532 nm picosecond laser pulses applied to 0.1 wt% zinc oxide microparticle suspension. Graph shows the dependence of laser fluence at the jet and laser fragmentation efficiency evaluated from the primary particle index.

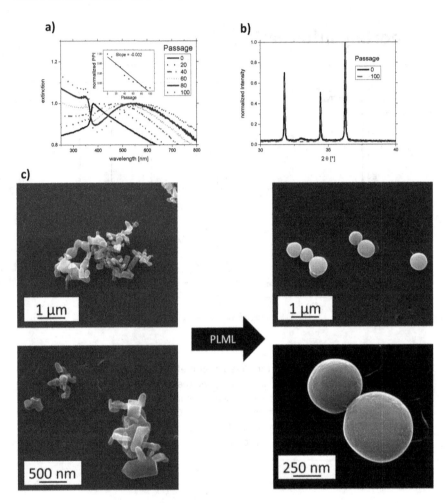

Figure 8-2:   Pulsed laser melting of crystalline zinc oxide micro particles at a laser
wavelength of 355 nm and 40 ns pulse duration. a) depicts the UV/Vis-
spectra and extracted PPI values for different number of passages of
the volume flow through irradiation zone. In b) the XRD-pattern before
and after 100 passages are shown. SEM pictures of the particles
corresponding to the XRD-pattern are shown in c) (left: before PLML,
right after PLML).

Due to the longer pulse duration (ns) thermal effects are much more distinctive and the shorter wavelength of 355 nm compared to 532 nm results in a higher absorption efficiency. Thus for 355 nm and nanosecond pulses whole ZnO agglomerates can melt down and form spheres with hydrodynamic diameters of around 400-500 nm. In case of picosecond pulses with 532 nm particles melt only partially forming smaller spheres or only partially melts down the surface, forming bigger sintering bridges of the aggregates.

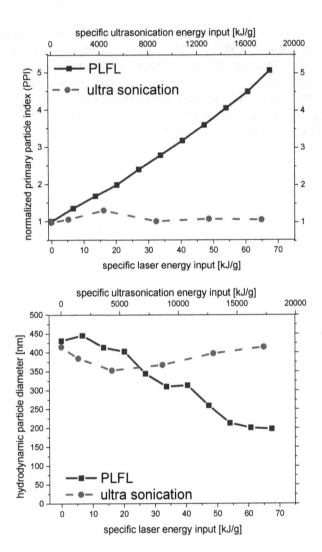

Figure 8-3:   Comparison of ultra-sonication and laser fragmentation. Top: primary particle index (PPI) and bottom: change of hydrodynamic particle diameter with increasing energy input.

Comparing plused laser fragmentation with strong ultra-sonication shows that PLFL is a top down method to fabricate des-aggregated primary particles and nanoparticles. A change of the crystallite size during PLFL can be observed by XRD analysis.

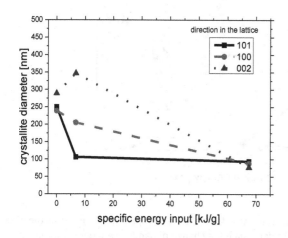

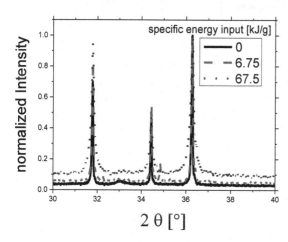

Figure 8-4:   Crystallite size (top) calculated from the XRD pattern (bottom) for the different directions in the wurtzite lattice of zinc oxide crystals for different laser energy input.

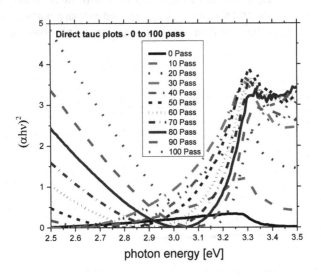

Figure 8-5:   Tauc plots calculated from the UV/Vis spectra $(\alpha h\nu)^2$ versus the photon energy) to determine the change of bandgap energy of the zinc oxide particles during PLFL.

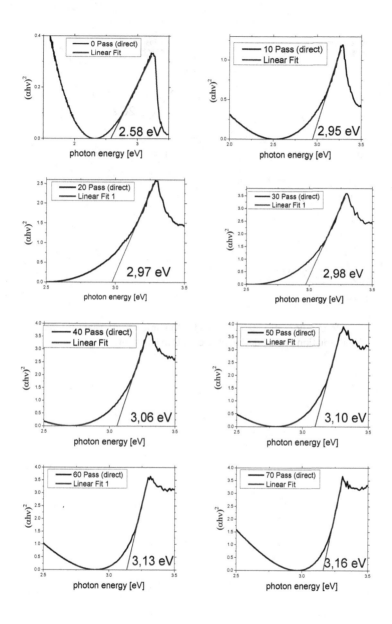

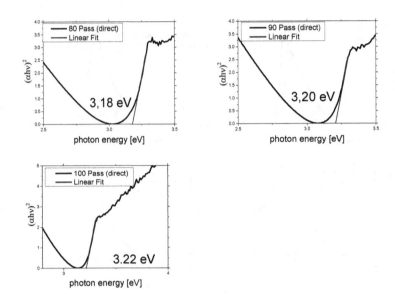

Figure 8-6:   Direct tauc plots with linear fits for the different passages (number of
irradiation cycles) during PLFL of ZnO in liquid jet.

Tab. 8-1:   Electron values determined from the direct tauc plots of Fig. 8-5 and
            8-6.

Number of passages	direct *eV*
0	2.58
10	2.81
20	2.86
30	2.89
40	3
50	3.03
60	3.08
70	3.12
80	3.14
90	3.17
100	3.22

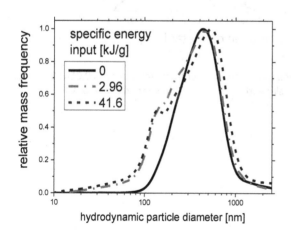

Figure 8-7:   Hydrodynamic particle size distributions of boron carbide for different
                       laser specific energy input. Note that no surfactant was used
                       subsequent to laser fragmentation and therefore nanoparticle

Supplementary Data 4.1.2

This diagram supports the statement that very small particles are generated for PLFL of ZnO which cannot be detected optically by UV-vis spectroscopy or ADC.

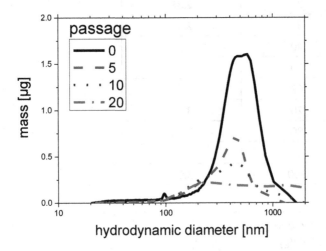

Figure 8-8:   Change of particle size for laser fragmentation with 0.01wt% zinc oxide microparticles with optimized fragmentation conditions.

Supplementary Data 4.1.3

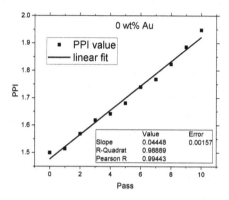

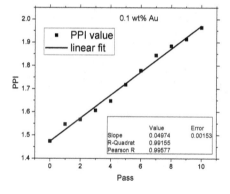

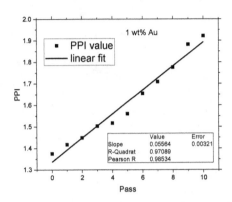

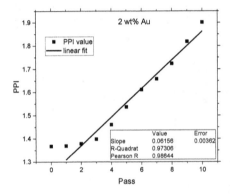

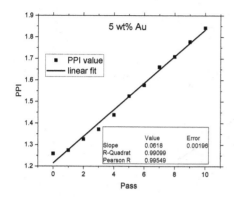

Figure 8-9:   PPI values plotted versus the number of passages for different gold
nanoparticle loadings on zinc oxide

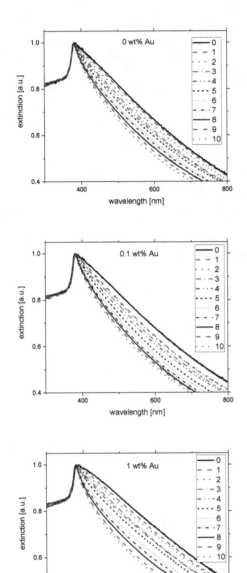

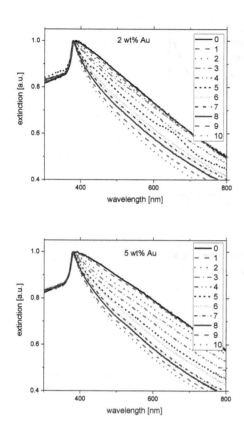

Figure 8-10:　　UV-vis spectra for the different gold nanoparticle loadings after each irradiation passage

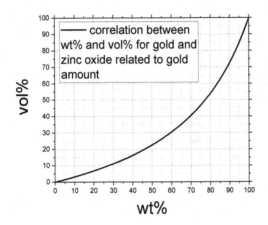

Figure 8-11:   Correlation between the weight and volume percentage for the
material composition gold/zinc oxide

Supplementary Data 4.1.4

M. Lau, I. Haxhiaj, P. Wagener, R. Intartaglia, F. Brandi, J. Nakamura, S. Barcikowski, *Ligand-free gold atom clusters adsorbed on graphene nano sheets generated by oxidative laser fragmentation in water*, Chem. Phys. Lett. 610 (**2014**) 256-260

1. Available number of reactive oxidative species

The figure S4.4 3 shows the calculated number of molecules in solution that are available for each surface atom of the particles at a gold concentration of 20 mg/L for water saturated with oxygen at 20°C and for a 10 wt-% hydrogen peroxide solution. For oxygen saturated water there are only few molecules available whereas hydrogen peroxide is present in an excess. Thus, hydrogen peroxide is quantitatively and qualitatively able to stabilize gold particles surface even at small particle sizes.

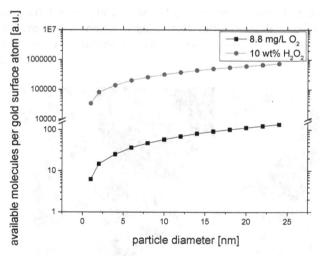

Figure 8-12:   Calculated number of molecules from reactive oxygen species available per surface atom in dependence of the particle size.

2. Redox potential

The potential shown in table 1 for a pH value of 5 and 8 was calculated by the Nernst equation:

$$E' = E^0 + \frac{RT}{zF}\ln(\frac{a_{ox}}{a_{red}})$$

Whereby E' is the calculated potential for a pH value of 5 or 8, E° is the standard potential for a pH value of 0, R the universal gas constant, T the temperature, F the Faraday constant, z the number of transferred electrons, a_{ox} the activity of the oxidant, and a_{red} the activity of the reductant. For the calculation of E' the values for E° were taken from literature [53].

3. TEM images of laser generated gold nanoparticles

Figure S4.4 4, S4.4 5 and S4.4 6 show representative TEM pictures that were used for the particle size characterization for the histograms in Figure 1 for gold nanoparticles after laser ablation, after laser fragmentation in presence of sodium hydroxide, and in presence of both sodium peroxide and hydrogen peroxide.

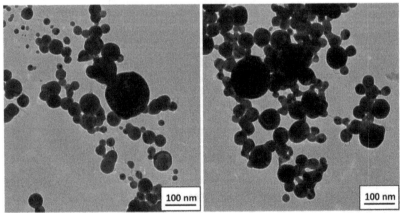

Figure 8-13:   Representative TEM pictures of the gold nanoparticles after laser ablation synthesis used for particle size characterization

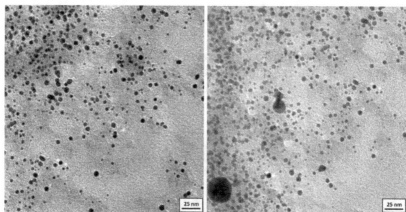

Figure 8-14:   Representative TEM pictures of the Au nanoparticles after laser fragmentation in presence of sodium hydroxide used for particle size characterization

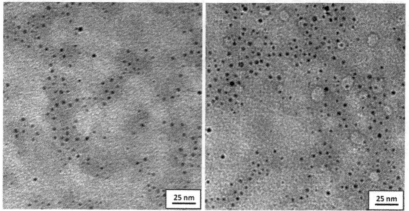

Figure 8-15:   Representative TEM pictures of the Au nanoparticles after laser fragmentation in presence of sodium hydroxide and hydrogen peroxide used for particle size characterization

4. HR-TEM of Au clusters and Au clusters on GNS

Figure S4.4 7 and S4.4 8 show some of the HR-TEM pictures that were used for the particle size characterization for the histograms in Figure 3 of gold nanoparticles before and after the support on GNS. Figure S4.4 7 include the electron diffraction pattern of a gold nanoparticle.

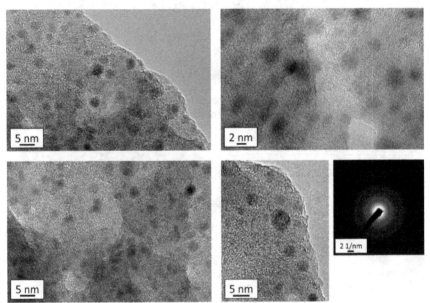

Figure 8-16:   Representative HR-TEM pictures of the Au clusters used for particle size characterization and diffraction pattern (bottom right)

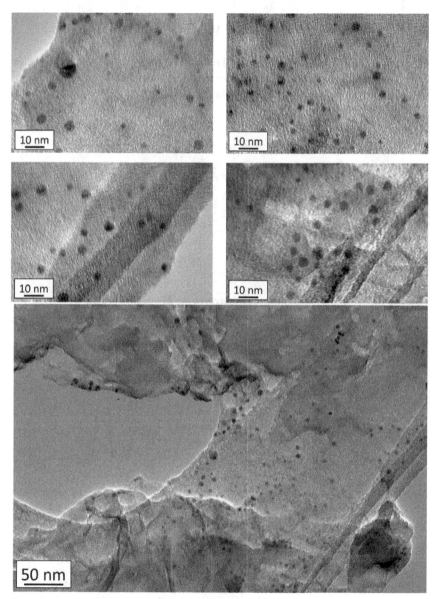

Figure 8-17:   Representative HR-TEM pictures of the Au clusters adsorbed on
                     graphene nano sheets (GNS)

5. Analytical disc centrifuge (ADC) analysis

Figure S7, S8 and S9 show the mass and number distribution of Au nanoparticles after laser ablation synthesis, after laser fragmentation with sodium hydroxide, and with sodium hydroxide and hydrogen peroxide.

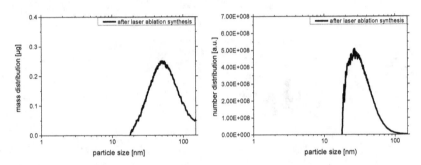

Figure 8-18:   Mass and number distribution of the Au nanoparticles after laser ablation synthesis measured by an analytical disc centrifuge

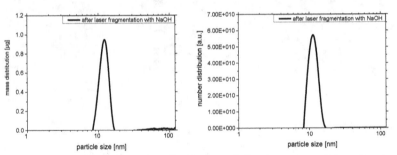

Figure 8-19:   Mass and number distribution of the Au nanoparticles after laser fragmentation in presence of sodium hydroxide measured by an analytical disc centrifuge

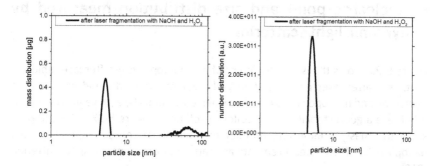

Figure 8-20:   Mass and number distribution of the Au nanoparticles after laser fragmentation in presence of sodium hydroxide and hydrogen peroxide measured by an analytical disc centrifuge

6. Isoelectric point and size distribution measured by dynamic light scattering

Figure 8-21 shows the isoelectric point of gold nanoparticles after laser ablation synthesis, after laser fragmentation with sodium hydroxide, and with sodium hydroxide and hydrogen peroxide. The isoelectric points by a pH value lower than 2 indicate a good stability of the colloidal solutions. Figure 8-22, 8-23 and 8-24 show the mass, volume and number distribution of the corresponding gold nanoparticles. These measurements are carried out by dynamic light scattering (DLS).

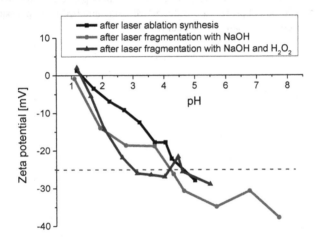

Figure 8-21:   Isoelectric point measurements of gold nanoparticles after laser ablation synthesis, after laser fragmentation with sodium hydroxide and with sodium hydroxide and hydrogen peroxide

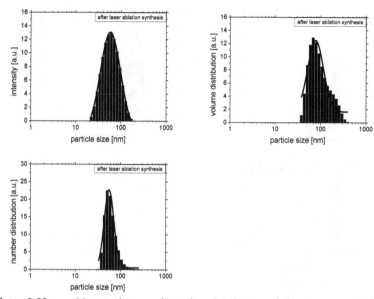

Figure 8-22: Mass, volume and number distribution of the Au nanoparticles after laser ablation synthesis measured by dynamic light scattering

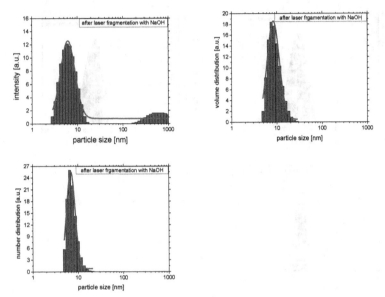

Figure 8-23:   Mass, volume and number distribution of the Au nanoparticles after laser fragmentation in presence of sodium hydroxide measured by dynamic light scattering

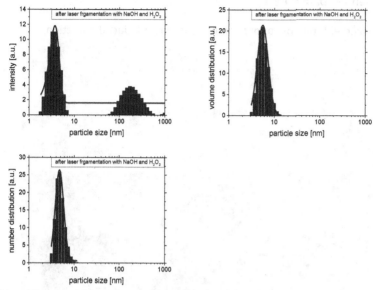

Figure 8-24: Mass, volume and number distribution of the Au nanoparticles after laser fragmentation in presence of sodium hydroxide and hydrogen peroxide measured by dynamic light scattering

Supplementary Data 4.1.5.1

4.1.5.1 Fragmentation of copper compounds - laser-induced reduction

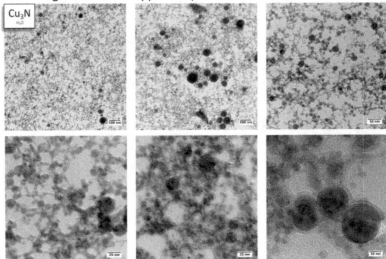

Figure 8-25: TEM images of nanoparticles generated from Cu₃N after laser fragmentation in water

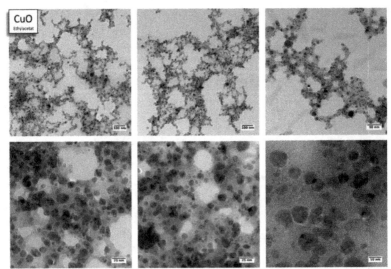

Figure 8-26: TEM images of nanoparticles generated from CuO after laser fragmentation in ethyl acetate

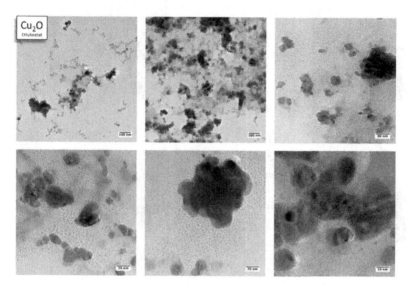

Figure 8-27:   TEM images of nanoparticles generated from Cu₂O after laser fragmentation in ethyl acetate

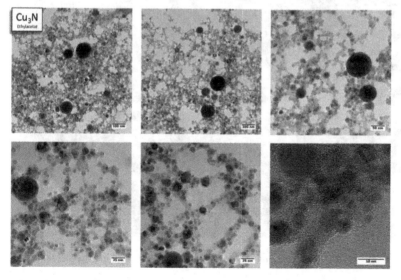

Figure 8-28:   TEM images of nanoparticles generated from Cu₃N in ethyl acetate

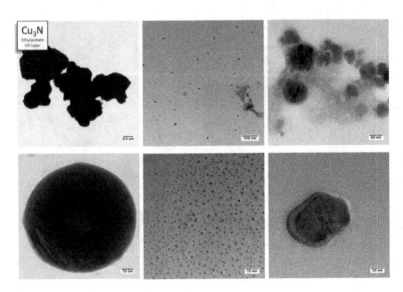

Figure 8-29:   TEM images of nanoparticles generated from Cu₃N in ethyl acetate by laser fragmentation with 355 nm

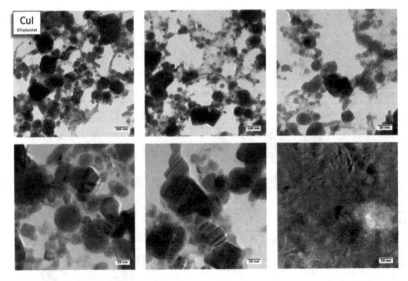

Figure 8-30:   TEM images of nanoparticles generated from CuI in ethyl acetate

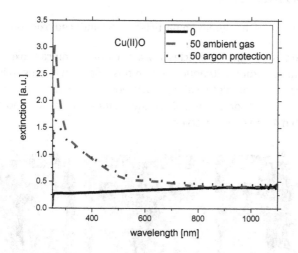

Figure 8-31: UV-vis spectra for the fragmentation of CuO in ethyl acetate with and without argon as protection gas

Supplementary Data 4.1.5.2

4.1.5.2 Fragmentation of aluminum particles - laser-induced oxidation

Laser activation of aluminum particles in water causes significant oxidation after a retention time of around 20 hours. During this oxidation hydrogen is released from the liquid and a complete white particle powder in water is obtained. From the reference material containing unirradiated aluminum particles no significant oxidation within 42 hours is observed.

Figure 8-32:   Photograph of aluminum particles suspended in water without laser treatment (left) and after 50 passages (right) for different storage times

Figure 8-33: Photograph of aluminum particles after 50 passages in different solvents after 66 hours storage

Supplementary Data 4.2.1

This supplementary data support chapter 4.2.1 on pulsed laser melting in liquids for zinc oxide microparticles and zinc oxide microparticles with supported gold nanoparticles. For the experimental studies fluence variation was applied by variation of liquid jets position to the focusing lens. Figure 8-34 shows the fluence resulting from the different positions. Hereby 0 means the focal plane of the 100 mm lens and 100 mm the position directly behind the lens.

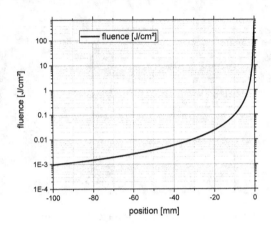

Figure 8-34:   Fluence for the 355 nm, 40 nm coherent laser used for laser melting studies with a 100 mm lens and a raw beam diameter of 6 mm, laser output power was 23 watt with 85 kHz and therefore a pulse energy of 270 µJ

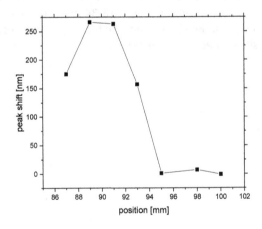

Figure 8-35:   Obtained peak shift from UV-vis spectra for the different processing positions for pure zinc oxide microparticles

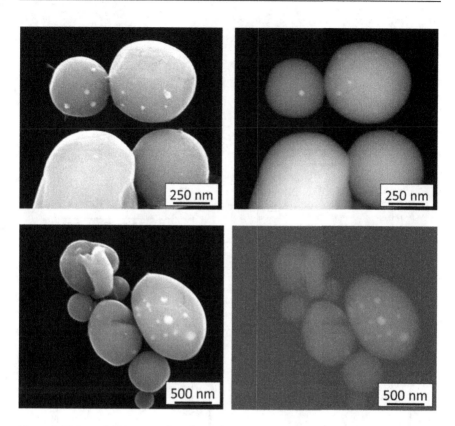

Figure 8-36: SEM images of zinc oxide mircoparticles after laser irradiation with 355 nm and 100 mJ/cm² detected with the secondary electron detector (left) and the back scattered electron detector (right)

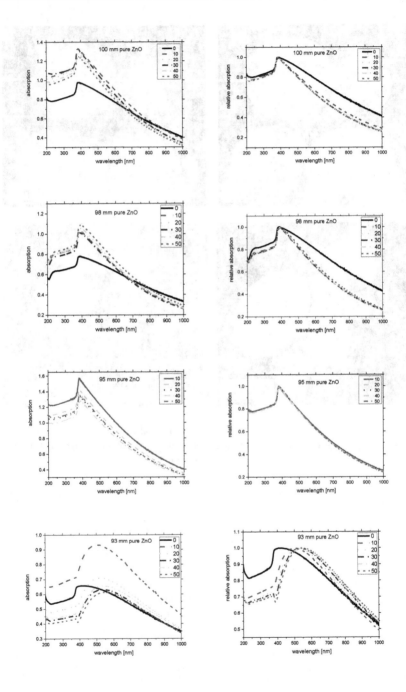

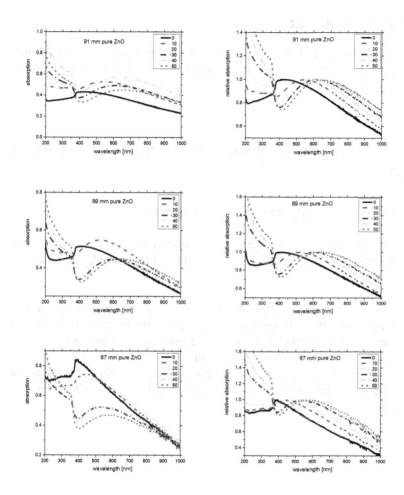

Figure 8-37: UV-vis spectra (not normalized left and normalized to the developed peak right) for the different processing positions after 10 passages respectively

Supplementary Data 4.2.2

M. Lau, A. Ziefuss, T. Komossa, S. Barcikowski, *Inclusion of supported gold nanoparticles into their semiconductor support*, Phys. Chem. Chem. Phys. 17 (**2015**) 29311-29318

Inclusion of supported gold nanoparticles into their semiconductor support

Marcus Lau, Anna Ziefuss, Tim Komossa, Stephan Barcikowski*

Technical Chemistry I, University of Duisburg-Essen and Center for Nanointegration Duisburg-Essen (CENIDE), Universitaetsstr. 7, 45141 Essen, Germany

These supporting information provide additional information on the properties of the materials. It is demonstrated how the particle mass load can be determined and a correlation between the mass and volume load is given. Further all UV-vis extinction spectra and additional size measurements determined by an analytical disc centrifuge are shown. Beside this, additional SEM images are provided, as the manuscript just shows a representative selection.

Supporting Information

a)

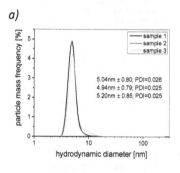

b)

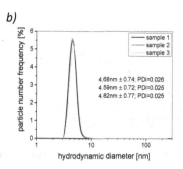

c)

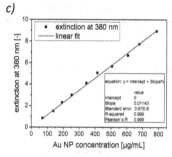

Figure 8-38:   Particle size distribution detected in an analytical disc centrifuge of
monodisperse Au NP colloids before adsorption onto zinc oxide support
a) as weight distribution and b) as number distribution of three different
samples; c) extinction at 380 nm for different Au NP colloid
concentrations and linear fit used to determine Au NP concentration
after centrifugation and prior to adsorption on zinc oxide

For calibration at an extinction wavelength of 380 nm laser ablation time was
varied from 1 to 10 minutes in 1 minute steps. The concentration of obtained
gold colloids was determined by differential weighing of the target before and
after laser ablation. The colloids were diluted to receive an extinction value
around 1 where the extinction scales linearly with the Au NP concentration.
Subsequently the measured extinction value was multiplied by the dilution factor
and plotted versus the concentration as shown in Fig. 8-38 c).

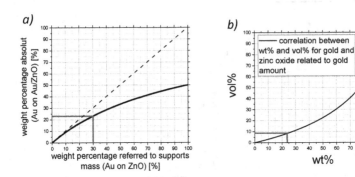

Figure 8-39:   a) correlation between the absolute weight percentage and the weight percentage referred to supports mass; b) correlation between volume percentage and weight percentage for the material composition of gold and zinc oxide referred to gold amount. Red lines mark the experimental values used for this study

Values for the curves shown in Fig. 8-39 a) and b) can be calculated by the following equations:

$$\frac{mass\ Au\ [g]}{mass\ ZnO\ [g]} \cdot 100 = weight\ percentage\ referred\ to\ support\ [wt\%]$$

$$\frac{mass\ Au\ [g]}{mass\ ZnO\ [g] + mass\ Au\ [g]} \cdot 100 = weight\ percentage\ absolute\ [wt\%]$$

$$\frac{\dfrac{mass\ Au\ [g]}{\rho_{Au}\ [\frac{g}{cm^3}]}}{\dfrac{mass\ ZnO\ [g]}{\rho_{ZnO}[\frac{g}{cm^3}]} + \dfrac{mass\ Au\ [g]}{\rho_{Au}\ [\frac{g}{cm^3}]}} \cdot 100 = volume\ percentage\ [vol\%]$$

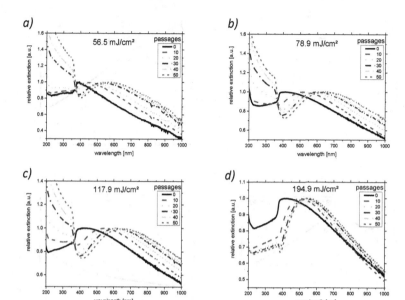

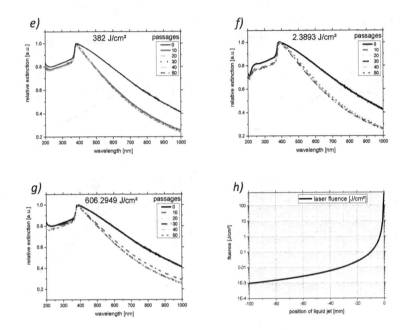

Figure 8-40:   UV/vis spectra of the pure zinc oxide particle suspensions for laser irradiation with 355 nm for increasing laser fluecnes (a-g) and calculated laser fluences for the different positions of the liquid jet relative to the focal plane of a 100 mm lens (h)

a)
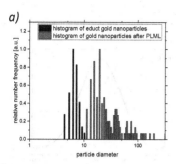

b)
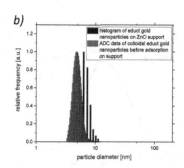

Figure 8-41:   Histogram of particle size distribution of gold nanoparticles on ZnO
support before (black bars) and included in ZnO SMS after PLML
derived from SEM images (red bars) (a)) and comparison of the Au NP
particle size distributions derived from analytical disc centrifuge (ADC,
red bars) before deposition onto ZnO and histogram from SEM images
after Au NP deposition (black bars) (b))

Fig. 8-41 shows that gold nanoparticle size does not change after deposition onto
the zinc oxide support but for PLML bigger gold nanoparticles are obtained. This
is in agreement with findings of Marzun et al. [Marzun, G.; Nakamura, J.; Zhang,
X.; Barcikowski, S.; Wagener, P. Size control and supporting of palladium
nanoparticles made by laser ablation in saline solution as a facile route to
heterogeneous catalysts. *Appl. Surf. Sci.* **2015**, *348*, 75-84] and Lau et al. [Lau, M.,
Haxhiaj, I.; Wagener, P.; Intartaglia, R.; Brandi, F.; Nakamura, J.; Barcikowski, S.
Ligand-free gold atom clusters adsorbed on graphene nano sheets generated by
oxidative laser fragmentation in water. *Chem. Phys. Lett.* **2014**, *610*, 256-260].
Further it demonstrates that high loadings can be achieved without causing
significant particle aggregation or agglomeration.

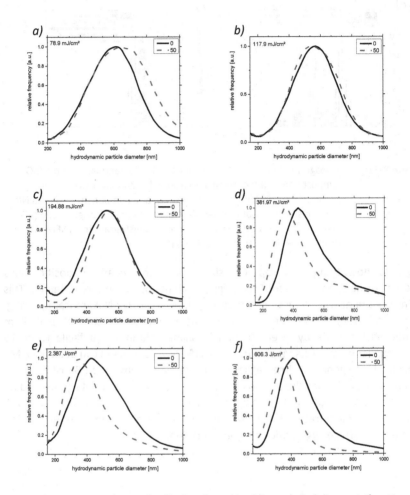

Figure 8-42:   Particle size distribution determined by analytical disc centrifugation
before and after 50 passages of pulsed laser melting of pure zinc oxide
in pure water for increasing laser fluences (a-f)

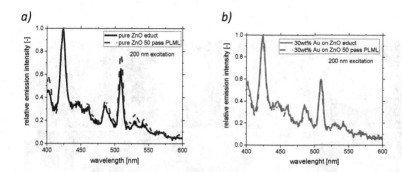

Figure 8-43: Fluorescence spectra of the pure ZnO before and after PLML a) and of 30 wt% Au NP ZnO particles before and after PLML b)

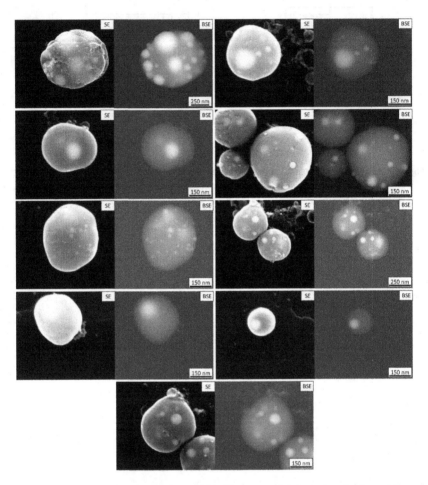

Figure 8-44:   Correlated SEM images of gold nanoparticle inclusions in zinc oxide
matrix forming sub-micrometer spheres after PLML with 79 mJ/cm²,
355 nm and 40 ns detected with a secondary electron detector (SE) and
back scattered electron detector (BSE)

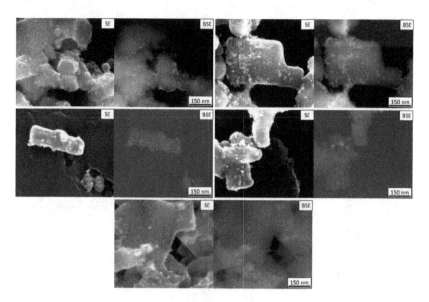

Figure 8-45:   Correlated SEM images of gold nanoparticles supported on zinc oxide particles before PLML detected with a secondary electron detector (SE) and back scattered electron detector (BSE)

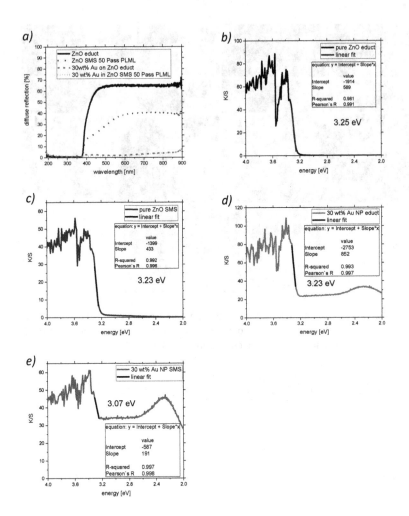

Figure 8-46:   a) diffuse reflection spectra for the different particle materials before and after 50 passages PLML; b)-e) determination of bandgap energy from linear fit by plotting the Kubelka-Munk function versus the electron energy for b) pure ZnO, c) ZnO SMS after 50 passages PLML, d) 30 wt% Au NP on ZnO and e) 30 wt% Au NP in ZnO SMS after 50 passages PLML

The ratio of absorption to scattering (K/S) can be determined by the following equation (Kubelka-Munk theory), with R_∞ as reflection of the infinite thick sample:

$$\frac{K}{S} = \frac{(1 - R_\infty)^2}{2R_\infty}$$

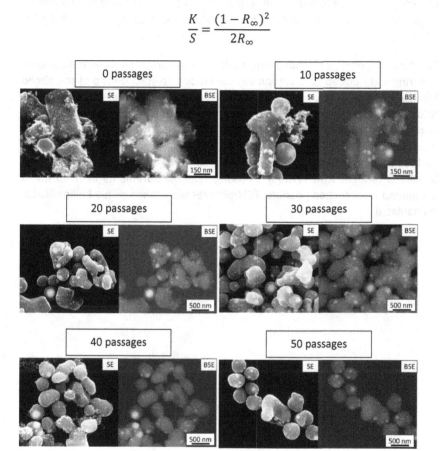

Figure 8-47: SEM images of particles after 0, 10, 20, 30, 40 and 50 passages of PLML with 30 wt% Au NP on ZnO detected with secondary electrons and back scattered electrons, respectively

Supplementary Data 4.2.3

M. Lau, R. Niemann, M. Bartsch, W. O'Neill, S. Barcikowski, *Near-field-enhanced, off-resonant laser sintering of semiconductor particles for additive manufacturing of dispersed Au–ZnO-micro/nano hybrid structures*, Appl. Phys. A 114 (**2014**) 1023-1030

Powders of 5 wt% gold-loaded zinc oxide microparticles before and after laser sintering were placed on a carbon sample holder. To stabilize top of the aimed lamella a small area (around 10x1 µm) was sputtered with gold. Afterwards holes beside the sputtered area are drilled, leaving a thin lamella of micro particles. Approaching a micro tip to the top of the particle wall to a few hundred nanometers, adhering it by sputtering and cutting the lamella from the powder bed by tilting carrier with particles allows to take out the lamella. Finally the taken lamella is brought to a TEM-grid, making TEM measurements of nano/micro compound cross section possible. TEM pictures were taken with a Philips CM12, Hamamatsu.

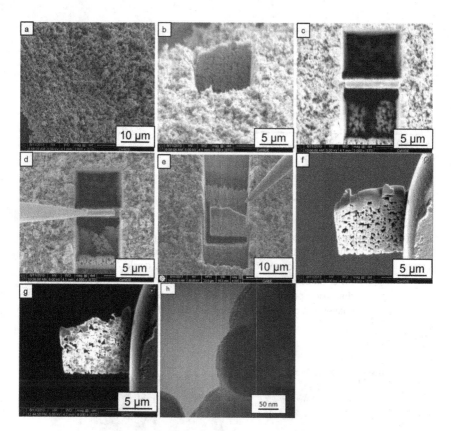

Figure 8-48:   SEM pictures of the FIB method to cut a powder bed of zinc oxide microparticles (a-g) for subsequent TEM analysis (h). In (a) an area on the supported micro particle powder is sputtered with gold. Afterwards a thin lamella is cut out by the ion beam (b-c). With approaching a micro tip and adhering the lamella, it can be taken out and glued to the edge of a TEM grid (f). To allow the observation of particles in TEM the lamella has to be thinned by focused ion beam (g). Transferring the lamella FIB cut enables examination of powder bed cross-section in a TEM (h).

■	no material response
▨	transition to sintering
▨	sintering
▨	transition to ablation
■	ablation

Figure 8-49:   Description of colors in sintering process parameter window diagrams out of optical microscope images corresponding to irradiated powder bed. The following diagrams and images depict sintering areas with variation of writing speed and laser power for different gold loading on zinc oxide powders at two different laser spot diameters. From these images and diagrams Figure 57 and 58 of the manuscript were prepared.

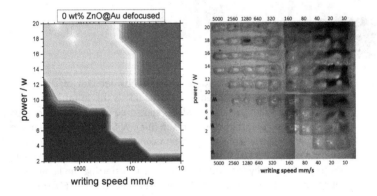

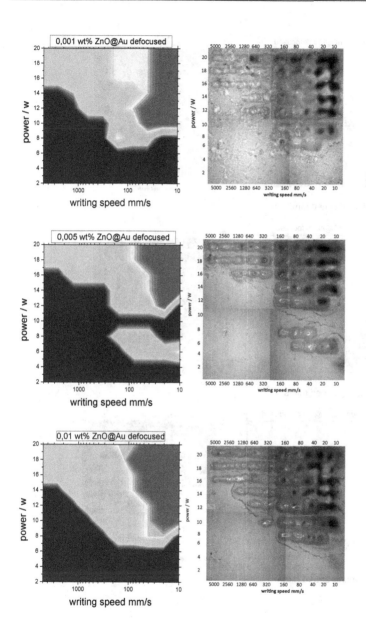

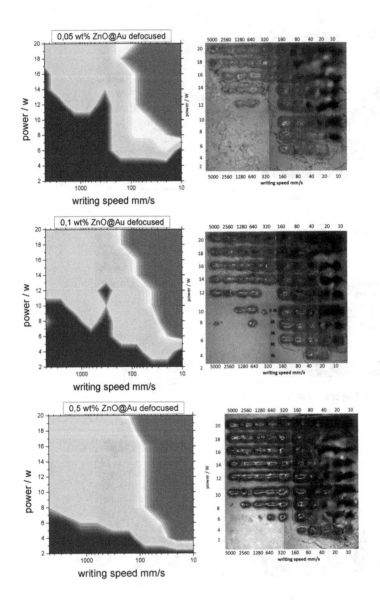

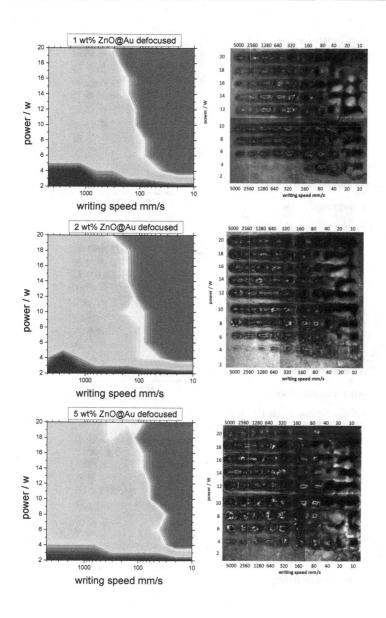

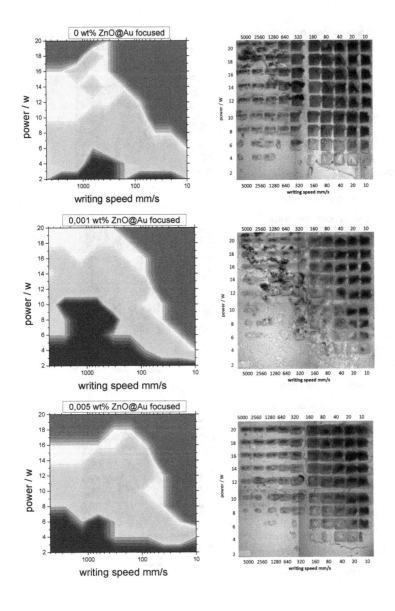

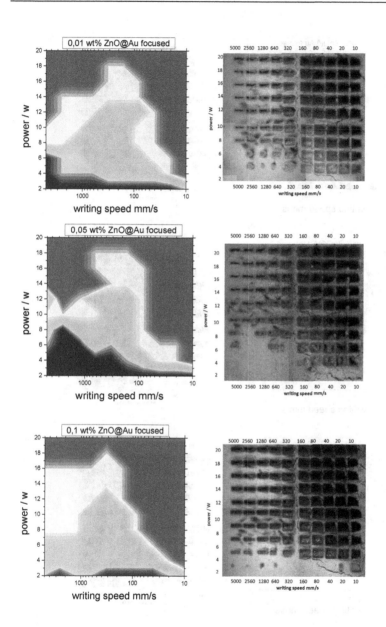

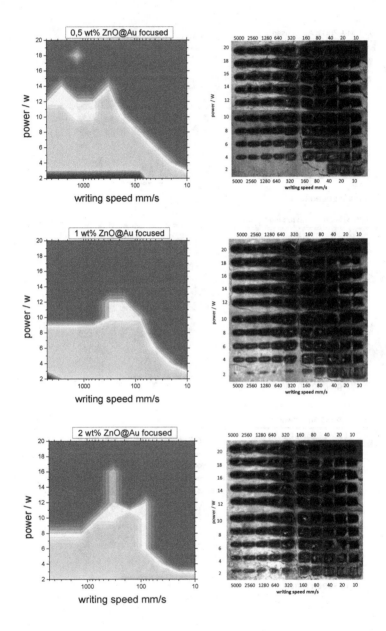

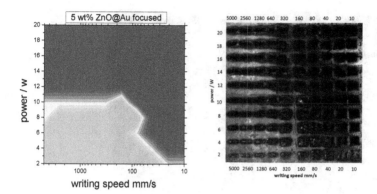

Supplementary Data 4.3

The following figure shows a cone impregnating a cylinder. Hereby the cone illustrates the laser light and the cylinder the liquid jet. To determine the ratio of illumination of the liquid jet by the laser beam it is assumed that no diffraction at the liquid-gas interphase occur. Due to geometrical reasons a section curve is formed at the surface of the liquid. This shape depends on the geometrical parameters of the cone and cylinder, respectively. Thus to determine the ratio of illumination with a formula an illuminated sphere segment at the front and back was assumed, as described in chapter 4.3.

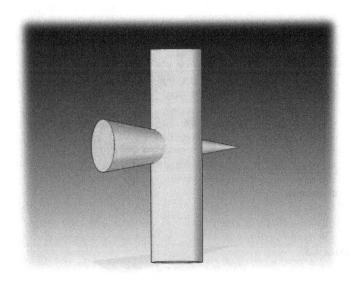

Figure 8-50:   Section of a cone in a cylinder forming the section curve at the surface of the liquid jet

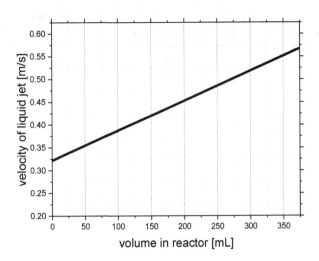

Figure 8-51:   Velocity of the liquid jet in dependence of the liquid volume in reactor

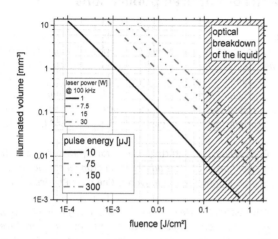

Figure 8-52: Illuminated volume plotted versus the fluence for different laser power assuming a liquid jet diameter of 1.3 mm, a laser raw beam diameter of 3.5 mm and a focal length of 100 mm of the lens

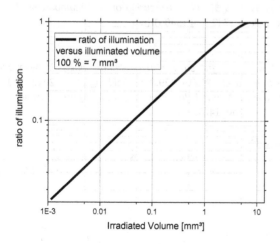

Figure 8-53: Correlation between the irradiated volume and the ratio of illumination for a 100 mm lens and a 1.3 mm liquid jet with a laser raw beam diameter of 3.5 mm

8.2 List of authored and co-authored publications

Peer-Reviewed-Publications:

1. **M. Lau**, A. Ziefuss, T. Komossa, S. Barcikowski, *Inclusion of supported gold nanoparticles into their semiconductor support*, Physical Chemistry Chemical Physics 17 (2015) 29311-29318
2. **M. Lau**, S. Barcikowski, *Quantification of mass-specific laser energy input converted into particle properties during picosecond pulsed laser fragmentation of zinc oxide and boron carbide in liquids*, Applied Surface Science 348 (2015) 22-29
3. **M. Lau**, I. Haxhiaj, P. Wagener, R. Intartaglia, F. Brandi, J. Nakamura, S. Barcikowski, *Ligand-free gold atom clusters adsorbed on graphene nano sheets generated by oxidative laser fragmentation in water*, Chemical Physics Letters 610 (2014) 256–260
4. **M. Lau**, R. Niemann, M. Bartsch, W. O'Neill, S. Barcikowski, *Near-field-enhanced, off-resonant laser sintering of semiconductor particles for additive manufacturing of dispersed Au–ZnO-micro/nano hybrid structures*, Applied Physics A 114 (2014) 1023-1030 (Invited Paper)
5. P. Wagener, **M. Lau**, S. Breitung-Faes, A. Kwade, S. Barcikowski, *Physical fabrication of colloidal ZnO nanoparticles combining wet-grinding and laser fragmentation*, Applied Physics A 108 (2012) 793-799
6. L. Jonušauskas, **M. Lau**, P. Gruber, B. Gökce, S. Barcikowski, M. Malinauskas, A. Ovsianikov, *Plasmon assisted 3D microstructuring of gold nanoparticle-doped polymers*, Nanotechnology 27 (2016) 154001 (8pp)

Patents:

1. **M. Lau**, S. Barcikowski, *Verfahren zur Herstellung reiner, insbesondere kohlenstofffreier Nanopartikel*, EP12194129.8 (Anmeldung 23. 11. 2012), (2013)
2. **M. Lau**, S. Barcikowski, *Method for manufacture of pure, carbon free nanoparticles*, US Patent App. 14/089,393, (2013)

Non-Peer-Reviewed-Publications:

1. **M. Lau**, B. Gökce, G. Marzun, C. Rehbock, S. Barcikowski, *Rapid nanointegration with laser-generated nanoparticles*. Lasers in Manufacturing 06/**2015**; Conference Proceedings

Oral Presentations (* presenting author):
1. **M. Lau***, S. Barcikowski, *Optical fabrication of defect-rich zinc oxide nanoparticles by laser fragmentation,* Nanofair, 2014, Dresden, Germany
2. **M. Lau***, R. Niemann, M. Bartsch, W. O'Neill, S. Barcikowski, *Amplified Additive Manufacturing by Gold Nanoparticles during Laser Sintering of Semiconductors,* LPM conference, 2014, Vilnius, Lithuania
3. **M. Lau***, R. Niemann, M. Bartsch, W. O'Neill, S. Barcikowski, *Near-Field-Enhanced, Off-Resonant Laser Sintering for Additive Manufacturing of Semiconductor Particles and the Fabrication of dispersed Au-ZnO-micro/nano hybrid structures,* Photonics Europe, 2014, Brüssel
4. **M. Lau***, S. Barcikowski, *Laser fragmentation and mass-specific energy balancing in a free liquid jet for fragmentation threshold determination,* 3rd ANGEL conference, 2014, Matsuyama, Japan
5. S. Barcikowski*, **M. Lau**, Rapid Nanoparticle Polymer Composites Prototyping by Laser Ablation in Liquids, 6th NRW Nano-Conference, 2014, Dortmund, Germany
6. **M. Lau***, S. Barcikowski, *Amplified laser fragmentation of microparticles in a free liquid jet,* 7th LAMP conference, 2015, Kitakyushu, Japan
7. **M. Lau**, B. Gökce, G. Marzun, C. Rehbock, S. Barcikowski*, *Rapid nanointegration with laser-generated nanoparticles.* Lasers in Manufacturing Conference, 2015, Munich, Germany
8. **M. Lau***, R. Niemann, M. Bartsch, W. O'Neill, S. Barcikowski, Hybridization of semiconductor micro particles with plasmonic nanoparticles during additive manufacturing, Lasers in Manufacturing Conference, 2015, Munich, Germany
9. I. Haxhiaj, **M. Lau***, G. Marzun, J. Nakamura, S. Barcikowski, *Ultra-small ligand-free noble metal nanoparticles produced by laser techniques for heterogeneous catalysis,* 6th Junges Chemie Symposium Ruhr, 2015, Essen, Germany
10. **M. Lau***, R. Niemann, M. Bartsch, W. O'Neill, S. Barcikowski, *Modifizierte generative Fertigung mit nanopartikelgedopten Trägermaterialien – Einfluss von Goldnanopartikel auf das Lasersintern von Zinkoxid,* JungakademikerInnen-Symposium Materialien für die Photonik, 2015, Essen, Germany
11. **M. Lau***, S. Barcikowski, *Functional Nanomaterials for Catalysis and Biomedicine,* Hannovermesse 2014, Hannover, Germany
12. **M. Lau***, S. Barcikowski, *Rapid Doping of Conventional Polymers with Laser-generated Nanoparticles for Optics, Magnetics, Energy and Biomedical Application,* Hannovermesse 2015, Hannover, Germany

Poster Presentations (* presenting author):

1. **M. Lau***, S. Barcikowski, *Laser Fragmentation for Nanoparticle Generation and Size Manipulation*. 3rd German-Japan Nanoworkshop, Tsukuba International Conference on Materials Science (TICMS), 2013, Tsukuba, Japan

2. **M. Lau***, P. Wagener, S. Breitung-Faes, A. Kwade, S. Barcikowski, *Mechanically-induced lattice defects in ZnO microparticles boost laser fragmentation efficiency*, 2nd ANGEL Conference, 2012, Taormina, Sicily

3. **M. Lau***, P. Wagener, S. Breitung-Faes, A. Kwade, S. Barcikowski, *Laser-assisted synthesis of ZnO nanoparticles*. ECIS - European Colloid and Interface Society; 2011, Berlin, Germany

4. **M. Lau***, K. Kujawski, S. Barcikowski, *Defect-rich zinc oxide nanoparticles by laser fragmentation*, Photonics Europe, 2014, Brussels, Belgium

5. **M. Lau***, I. Haxhiaj, P. Wagener, S. Barcikowski, *Change of Particle Properties by Laser Irradiation for Catalytic Applications*, CENIDE Nanobiophotonics symposium 2015, Essen, Germany

6. I. Haxhiaj*, **M. Lau**, R. Intartaglia, F. Brandi, J. Nakamura, S. Barcikowski, *Oxidative laser fragmentation for ligand-free non-plasmonic gold atom and their deposition on graphene nanosheets*, 3rd ANGEL conference, 2014, Matsuyama, Japan

7. I. Haxhiaj*, **M. Lau**, J. Nakamura , P. Wagener, S. Barcikowski, *Ultra-small ligand-free gold atom clusters generated by oxidative laser fragmentation in water and supported on graphene nanosheets*, 6th NRW Nano-Conference, 2014, Dortmund, Germany

8. I. Haxhiaj*, **M. Lau**, P. Wagener, J. Nakamura, S. Barcikowski, *Oxidative laser fragmentation in liquids – supplying ligand-free ultra-small gold clusters for potential application in catalysis*, Gordon research conference (GRC), 2015, Ventrua, California

9. I. Haxhiaj*, **M. Lau**, Y. Akasu, J. Nakamura, P. Wagener, S. Barcikowski, *Generation of gold nanoparticle/graphene nano sheets material compounds*, CENIDE-CNMM-TIMS Joint Symposium on Nanoscience and –technology, 2015, Duisburg, Germany

10. I. Haxhiaj, **M. Lau***, J. Nakamura, P. Wagener, S. Barcikowski, *Oxidative laser fragmentation for ultra-small pure gold clusters with subsequent support to graphene nanosheets*, 7th LAMP conference, 2015, Kitakyushu, Japan

11. I. Haxhiaj, **M. Lau***, G. Marzun, J. Nakamura, S. Barcikowski, *Ultra-small ligand-free noble metal nanoparticles produced by laser techniques for heterogeneous catalysis*, 6th Junges Chemie Symposium Ruhr, 2015, Essen, Germany

12. **M. Lau***, L. Jonušauskas, M. Malinauskas, P. Gruber, A. Ovsianikov, S. Barcikowski, *Laser-generated nanoparticles for 3D additive manufacturing*, JungakademikerInnen-Symposium Materialien für die Photonik, 2015, Essen, Germany

13. **M. Lau***, P. Wagener, C. Rehbock, B. Gökce, S. Barcikowski, *Rapid Nanointegration*, 6th NRW Nano-Conference, 2014, Dortmund, Germany

Scientific Videos with Creative Commons Licenses:

1. S. Kohsakowski, **M. Lau**, S. Reichenberger, G. Marzun, P. Wagener, S. Barcikowski, *Catalyst for Water Splitting Hydrogen Gas derived from Water - Made by Laser Ablation in Pure Water*, 18.02.2014, https://www.youtube.com/watch?v=rMedEbJaEO4

2. S. Kohsakowski, **M. Lau**, S. Reichenberger, G. Marzun, P. Wagener, S. Barcikowski, *Photocatalytic Water Splitting Catalyst Fabrication by Nanoparticle Adsorption on Titatium Dioxide*, 18.02.2014, https://www.youtube.com/watch?v=o3wBcq4Cbtl

3. **M. Lau**, J. Jakobi, S. Barcikowski, *Nanosilver without Precursor Chemistry: Realtime Silver Nanoparticle Formation During Laser Ablation*, 27.11.2012, https://www.youtube.com/watch?v=ITTs7V6Yhws

4. **M. Lau**, J. Jakobi, S. Barcikowski, *nanoparticle agglomerate size reduction and fragmentation by laser excitation*, 02.02.2012, https://www.youtube.com/watch?v=4n1AoN2YNzo

5. **M. Lau**, J. Jakobi, S. Barcikowski, *clean nanoparticle colloids for medical and catalysis application*, 02.02.2012, https://www.youtube.com/watch?v=J0vZgQFDyjo

6. **M. Lau**, S. Barcikowski, *Quantification of mass-specific laser energy input converted into particle properties during picosecond pulsed laser fragmentation of zinc oxide and boron carbide in liquids*, Audio Slides presentation, Applied Surface Science, doi: 10.1016/j.apsusc.2014.07.053, 24.02.2015, http://audioslides.elsevier.com/ViewerSmall.aspx?doi=10.1016/j.apsusc.2014.07.053&Source=0&resumeTime=0&resumeSlideIndex=0&width=800&height=63

Awards:

2015   2nd place for the presentation „*Modifizierte generative Fertigung mit nanopartikelgedopten Trägermaterialien – Einfluss von Goldnanopartikel auf das Lasersintern von Zinkoxid*" (Oral presentation at the JungakademikerInnen-Symposium Materialien für die Photonik, 2015, Essen, Germany)

The LPM 2015 Outstanding Student Paper Award (Oral) for the presentation "*Amplified laser fragmentation of microparticles in a free liquid jet*" (Oral presentation at LAMP 2015)

2014   CENIDE best paper award 2014 for outstanding publication for the paper „*Ligand-free gold atom clusters adsorbed on graphene nanosheets generated by oxidative laser fragmentation in water*", Chemical Physics Letters 610, 256 (2014)

LPM 2014 outstanding student paper award (oral) for the presentation "*Amplified Additive Manufacturing by Gold Nanoparticles during Laser Sintering of Semiconductors*" (Oral presentation at LPM 2014) (Silver)

2007   1st place at the "*VDI-Aufwindkraftwerkwettbewerb 2007*" by the Institute of Heat and Fuel Technologies, Braunschweig, July 2007

8.3 Declaration

Parts of this work have been published, with my own contribution declared as follows:

- Chapter 4.1.1 has been published in Applied Surface Science [Lau2014a]: M. Lau, S. Barcikowski, *Quantification of mass-specific laser energy input converted into particle properties during picosecond pulsed laser fragmentation of zinc oxide and boron carbide in liquids*, Appl. Surf. Sci. 348 (**2015**) 22-29

    o Declaration of own contribution: The experimental design, experiments, characterization, analysis of results, data and draft of the manuscript were done by ML. This was conducted under the supervision of SB who also revised the manuscript.

- Chapter 4.1.4 has been published in Chemical Physics Letters [Lau2014d]: M. Lau, I. Haxhiaj, P. Wagener, R. Intartaglia, F. Brandi, J. Nakamura, S. Barcikowski, *Ligand-free gold atom clusters adsorbed on graphene nano sheets generated by oxidative laser fragmentation in water*, Chem. Phys. Lett. 610 (**2014**) 256-260

    o Declaration of own contribution: Preliminary and initial experiments, transfer and analysis of gold clusters on GNS, determination of oxidative potentials and draft of the manuscript were done by ML. This was conducted under the supervision of SB who also revised the manuscript. IH conducted detailed experimental studies, contributed several Figures and improved the manuscript during several internal reviews. PW, RI, FB and JN reviewed the manuscript and improved the English.

- Chapter 4.2.2 has been published in Physical Chemistry Chemical Physics [Lau2015], M. Lau, A. Ziefuss, T. Komossa, S. Barcikowski, *Inclusion of supported gold nanoparticles into their semiconductor support*, Phys. Chem. Chem. Phys. 17 (**2015**) 29311-29318

  o Declaration of own contribution: Initial melting experiments, data analysis, figures and draft if the manuscript were done by ML. This was conducted under the supervision of SB who also revised the manuscript. AZ prepared monodisperse gold particles and supported them onto the zinc oxide. TK conducted fluence studies on the laser melting of the pure zinc oxide.

- Chapter 4.2.3 has been published in Applied Physics A [Lau2014e]: M. Lau, R. Niemann, M. Bartsch, W. O'Neill, S. Barcikowski, *Near-field-enhanced, off-resonant laser sintering of semiconductor particles for additive manufacturing of dispersed Au–ZnO-micro/nano hybrid structures*, Appl. Phys. A 114 (**2014**) 1023-1030

  o Declaration of own contribution: Evaluation, analysis, interpretation of data and construction of the manuscript (including calculations and graphical illustrations) was realized by ML. This was done under the supervision of SB who also revised the manuscript. RN prepared the gold nanoparticles supported on zinc oxide and made experimental laser studies of the supported particles under the supervision of ML. MB operated the focused ion beam and cut the lamella for further analysis. WO supervised experimental work at the University of Cambridge.

Printed in the United States
By Bookmasters